Molecular Epidemiology Study of Mycobacterium Tuberculosis Complex

Edited by Yogendra Shah

Published in London, United Kingdom

IntechOpen

Supporting open minds since 2005

Molecular Epidemiology Study of Mycobacterium Tuberculosis Complex
http://dx.doi.org/10.5772/intechopen.92916
Edited by Yogendra Shah

Contributors
Dinara Namazovna Adjablaeva, Krzysztof Anusz, Anna Didkowska, Monika Krajewska-Wędzina, Blanka Orłowska, Monika Kozińska, Ewa Augustynowicz-Kopeć, Rabie Ayari, Ramy Triki, Youssef Mallat, Achraf Abdennadher, Khalil Amri, Raja Amri, Mohamed Ali Sbai, Mohammad Asif, Saad Alghamdi, Onix Cantres-Fonseca, Francisco Del Olmo, Manish Dwivedi, Priya Giri, Yogendra Shah

First published in London, United Kingdom, 2021 by IntechOpen
IntechOpen is the global imprint of INTECHOPEN LIMITED, registered in England and Wales, registration number: 11086078, 5 Princes Gate Court, London, SW7 2QJ, United Kingdom
Printed in Croatia

British Library Cataloguing-in-Publication Data
A catalogue record for this book is available from the British Library

Additional hard and PDF copies can be obtained from orders@intechopen.com

Molecular Epidemiology Study of Mycobacterium Tuberculosis Complex
Edited by Yogendra Shah
p. cm.
Print ISBN 978-1-83968-099-1
Online ISBN 978-1-83968-100-4
eBook (PDF) ISBN 978-1-83968-101-1

For EU product safety concerns:
IN TECH d.o.o., Prolaz Marije Krucifikse Kozulić 3, 51000 Rijeka, Croatia, info@intechopen.com or visit our website at intechopen.com.

Meet the editor

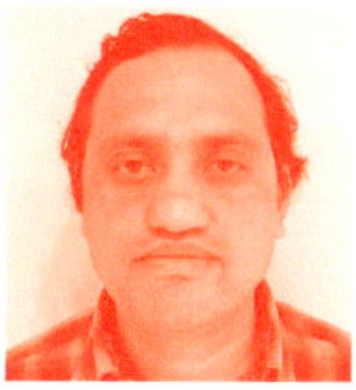

Dr. Yogendra Shah is a consultant microbiologist/virologist, senior research microbiologist, and lecturer at Seti Provincial Hospital, COVID-19 PCR laboratory, National Zoonoses and Food Hygiene Research Center, and Kathmandu College of Science and Technology, Nepal. He obtained a Ph.D. in Veterinary Medicine (Bacteriology) from the Graduate School of Veterinary Medicine, Hokkaido University, Japan, in 2017. His research focuses on better understanding the molecular epidemiological features/transmission dynamics of infectious diseases and zoonotic infectious diseases in Nepal by employing molecular techniques like ELISA, polymerase chain reaction (PCR), loop-mediated isothermal amplification (LAMP), and DNA sequencing. He was awarded the Young Science and Technology Award from the Nepal Academy of Science and Technology (NAST) in 2019. His research interests include infectious diseases, zoonotic infectious diseases, and vector-borne diseases. He has published more than thirty-five research articles in peer-reviewed journals and twelve books.

Contents

Preface

Tuberculosis (TB), which is caused by the infectious agent *Mycobacterium tuberculosis* (MTB), is a major global public health problem that infects one third of the world's population. However, TB is a curable and preventable disease. The World Health Organization (WHO) estimates that 10 million new cases of TB and 1.2 million deaths due to TB occurred in 2019. Globally, in 2019, close to half a million people developed rifampicin-resistant TB (RR-TB), 78% of whom had multidrug-resistant TB (MDR-TB). Molecular genotyping of MTB has been well developed over the years. This book provides an overview of the molecular epidemiology pattern, transmission dynamics, host response, evolution, and pathogenesis mechanisms of TB. It also examines mechanisms associated with increasing trends of drug-resistant TB and explores the development of anti-mycobacterial drugs. It presents updated research for policymakers and planners on diagnosis and treatment, genotyping tools, and control and prevention of MTB disease. The book also explores vigorous approaches in designing novel anti-tubercular drugs, diagnosis and treatment of latent tuberculosis infection, laboratory diagnosis via identifying novel single-nucleotide polymorphism, tracing of outbreak isolates, molecular characterization of *Mycobacterium* spp. isolated from cattle and wildlife in Poland, challenges in drug discovery against tuberculosis, and genealogy of resistant TB in Latin American territories.

Yogendra Shah
COVID-19 PCR Laboratory and Seti Provincial Hospital,
Dhangadhi, Nepal

Section 1

Introduction

Chapter 1

Introductory Chapter: Plan to Prevent and Combat against the Drug-Resistant Tuberculosis/ Zoonotic Tuberculosis

Yogendra Shah

1. Molecular Epidemiology of MTB drug resistant

Tuberculosis (TB) is a primary cause of death from a single infectious agent by *Mycobacterium tuberculosis* complex (MTBC) remains a major global public health problem which infects one thirds of world's population. Despite being largely TB is a curable and preventable disease, WHO estimates that 10 million new cases and 1.2 million deaths occurred in 2019 [1]. Majority of deaths were in developing countries with more than half occurring in Asia and Africa. TB is spread when people who are sick with TB expel bacteria into the air; for example, by coughing. TB usually affects the lungs (pulmonary TB), although other organs are involved in 15–30% of other sites (extra pulmonary TB) [2]. *Mycobacterium tuberculosis* (MTC or MTBC) is a genetically related group of *Mycobacterium* species that can cause tuberculosis in human or other animals i.e. *M. tuberculosis*, *M. africanum*, *M. orygis*, *M. bovis*, *M. microti*, *M. canetti*, *M. caprae*, *M. pinnipedii*, *M. suricattae*, *M. mungi* [3].

The emergence of drug resistant including multi-drug resistant (MDR-TB: It means that the TB bacteria that a person is infected with are resistant to two of the most important TB drugs, isoniazid (INH) and rifampicin (RMP) [4] and Extensively drug-resistant TB (XDR-TB) is a rare type of multidrug-resistant tuberculosis (MDR TB) that is resistant to isoniazid and rifampin, plus any fluoroquinolone and at least one of three injectable second-line drugs (i.e., amikacin, kanamycin, or capreomycin) [5] are also poses a serious public health threat to the success of TB treatment and control programs across worldwide. Globally in 2019, close to half a million people developed rifampicin-resistant TB (RR-TB), of which 78% had multidrug-resistant TB (MDR-TB). Molecular genotyping of MTB has been well developed over the years. WHO has developed the End TB strategy, which was endorsed by the sixty-seventh world health assembly in 2014. According to WHO, strategy ambitiously proposes to "end the global TB epidemic" by 2035. The strategy targets a 90% reduction in patients suffering from TB, and a 95% reduction in deaths from TB by 2035-all while protecting families from catastrophic costs that push them further into poverty [6].

The main genotyping typing methods mainly IS6110 restriction fragment length polymorphism (RFLP), spoligotyping and mycobacterial interspersed repeat unit variable-number tandem repeat (MIRU-VNTR) analysis, are commonly used for fingerprinting MTB strains to detect recent transmission [7]. However, the discriminatory power of these genotyping methods is not sufficient in countries such as South East Asia, South Africa and Russia including Nepal where MTB of Beijing family has been reported in high prevalence [8]. These factors might be the driving force for the spreading and emergence of MDR-TB as well as extensively drug resistant TB (XDR-TB) involved in clonal expansion of strains [9]. Even though, molecular genotyping techniques have been developed, they provide less discriminatory power to differentiate the genetic diversity, transmission dynamics and outbreak of MTB strains. Even in clustered isolates these methods could not distinguish the recent from past transmissions [10]. Furthermore, genomic heterogeneity among the drug susceptible or drug resistant strains could also not be accurately detected using conventional genotyping methods [11]. Whole-genome sequencing (WGS) based on next-generation sequencing (NGS) has been emerging as a very powerful tools for detection of genetic diversity, outbreak analysis, surveillance and determination of drug resistance [12]. Recently, WGS is considered as a gold standard method because of its high resolution allowing for in-depth characterization about the dynamics of evolution, transmission and exogenous infection [13].

The main importance of this book chapter was to provide overview and also understand about the molecular epidemiology pattern, transmission dynamics, host response, mechanisms associated with increasing trends of drug resistant TB including MDR-TB, evolution, molecular biology, pathogenesis mechanism and development of anti-mycobacterial drugs about the *Mycobacterium tuberculosis* complex. The purpose of book chapter will be help to provide the updating research information to the policy maker or planner for further diagnosis and treatment with genotyping tools, control and prevention for MTB disease. This book chapter main theme are to explore the vigorous approaches in novel designing of anti-tubercular drugs, diagnosis and treatment of latent tuberculosis infection to measure their quality of life, laboratory diagnosis by identification of novel SNPs, tracing of outbreak isolates, study of various chemically and structurally diverse currently clinically used and recently developed for anti-mycobacterial drugs, molecular characterization of *Mycobacterium* spp. isolated from cattle and wildlife in Poland, challenges in drug discovery against tuberculosis and genealogy of resistant TB in Latin American territories.

Author details

Yogendra Shah
Seti Provincial Hospital, COVID-19 PCR Lab, Dhangadhi, Kailali, Nepal

*Address all correspondence to: yogendra90@hotmail.com

References

[1] World Health Organization. Global tuberculosis report 2020. https://apps.who.int/iris/bitstream/handle/10665/336069/978924001313.

[2] Transmission and oathogeneis of tuberculosis https://www.cdc.gov/tb/education/corecurr/pdf/chapter2.pdf

[3] Jane ES, Gunn-Moore DA. Mycobacterial Infections. Canine and feline Infectious Diseases. 2014.

[4] Tuberculosis: Multidrug-resistant tuberculosis (MDR-TB) https://www.who.int/news-room/q-a-detail/tuberculosis-multidrug-resistant-tuberculosis-(mdr-tb)

[5] Tuberculosis https://www.cdc.gov/tb/publications/factsheets/drtb/xdrtb.htm

[6] https://www.who.int/westernpacific/activities/implementing-the-end-tb-strategy

[7] Barnes PF, Cave MD. Molecular epidemiology of tuberculosis. The New England Journal of medicine 2003; 349(12):1149-1156.

[8] Allix-Beguec C, Wahl C, Hanekom M, Nikolayevskyy et al. proposal of a consensus set of hypervariable mycobacterial interspersed repetitive-unit-variable-number tandem repeat loci for subtyping of *Mycobacterium tuberculosis* Beijing isolates. Journal of clinical microbiology 2014; 52(1):164-172.

[9] Iwamoto T, Grandjean L, Arikawa K, Nakanishi N, Caviedes L, Coronel J, et al. Genetic diversity and transmission characteristics of Beijing family strains of *Mycobacterium tuberculosis* in Peru. PLoS One 2012;7.

[10] Walker TM, Ip CL, Harrell RH, Evans JT, Kapatai G, Dedicoat MJ, et al. 2013. Whole-genome sequencing to delineate Mycobacterium tuberculosis outbreaks: A retrospective observational study. The Lancet Infectious diseases 13(2):137-146.

[11] Niemann S, Koser CU, Gagneux S, Plinke C, Homolka S, Bignell H, et al. 2009. Genomic diversity among drug sensitive and multidrug resistant isolates of Mycobacterium tuberculosis with identical DNA fingerprints. PloS one 4(10):e7407.

[12] Hasnain SE, O'Toole RF, Grover S, Ehtesham NZ 2015.Whole genome sequencing: A new paradigm in the surveillance and control of human tuberculosis. Tuberculosis (Edinb) 95(2):91-94.

[13] Kato-Maeda M, Metcalfe JZ, Flores L 2011. Genotyping of Mycobacterium tuberculosis: Application in epidemiologic studies. Future Microbiol 6(2): 203-216

Section 2

Drug Resistant TB and Challenges in Drug Discovery

Chapter 2

Challenges in Drug Discovery against Tuberculosis

Manish Dwivedi and Priya Giri

Abstract

Tuberculosis (TB) is one of the deadly diseases in the present era caused by *Mycobacterium tuberculosis*. Principally, this bacterium attacks the lungs, however, MTB Has been observed affecting any part of the human body including the kidney, spine, and brain. Drug-resistant progression and other associated properties of MTB become a major hurdle in drug discovery to fight against tuberculosis. Moreover, some of the challenging situations such as the low range of chemical agents, the time-consuming process of drug development, the shortage of predictive animal models, and inadequate information of the physicochemical evidence required for effective bacterial penetration, are additional hindrances for the pharmaceutical scientist. In the current chapter, we focus on challenges encountered during drug discovery and need to be overcome as *M. tuberculosis* has a substantial barrier in its lipid-containing cell wall to inhibit the influx of drugs which is the initial requirement of the drug to show its therapeutic effect. There is also an immediate need for efficient vaccine development which may show its effect on adolescents and adults along with infants. Investigation on key bacterial targets has been troublesome, in light of the vulnerability around the microenvironments found in vivo and subsequently, the importance of exceptional metabolic pathways. The manuscript is prepared after the extensive literature survey to explore the vigorous approaches in novel drug designing and in proposing potent drug targets. The re-engineering and repositioning of prominent antitubercular drugs are required to attain viable control.

Keywords: *Mycobacterium tuberculosis*, drug, challenges, bacterial targets

1. Introduction

Tuberculosis (TB), one of the most common deadly disease is caused by a bacterium called *Mycobacterium tuberculosis*. Robert Koch in 1882, isolated the mammalian strain and proved that the *Mycobacterium tuberculosis* plays a causative role in Tuberculosis. As per the latest WHO report approximately one-fourth of the world's population are infected with *Mycobacterium tuberculosis* (Mtb), whereas 5–10% of the total will develop TB disease during their lifetime [1, 2]. The WHO estimated that in 2018, about 10 million people were affected due to TB worldwide and 1.5 million people suffering from the ailment, including 2,51,000 people who additionally had HIV [3, 4]. In the past, TB was a major reason for death around the globe [5, 6]. In industrialized nations, TB is getting slow due to vast development and improvements in drugs and new antibiotics [5, 7].

TB may exist in two forms, active (dynamic) TB and Latent TB. Dynamic tuberculosis is a condition where MTB causes contamination; regularly, in the lungs, albeit numerous frameworks can be included. Dynamic TB is a multiorgan illness brought about by essential disease or as reactivation of inert tuberculosis. As need be, dynamic tuberculosis could be essential tuberculosis or reactivation tuberculosis.

Latent TB happens when an individual has the TB microscopic organisms inside their body, however, the microbes are available in tiny numbers. They are monitored by the body's safe framework and do not bring on any indications. Individuals with idle TB do not feel wiped out and are not irresistible. They cannot give the TB microscopic organisms to others. Moreover, they will generally have an ordinary chest x-ray and a negative sputum test. It is regularly just realized that somebody has latent TB since they have had a TB test, for example, the TB skin test. There are two kinds of test that can be utilized. These are the TB skin test (TST) and the fresher IGRA blood test. In nations where there is a significant degree of TB, (for example, the high weight TB nations) most individuals may have latent TB.

Fortunately, most of the TB patients have latent infection i.e., bacteria are present in the body but is not causing active disease. Hence at any one time, there are about 10 million people across the world with active tuberculosis infection and that causes deaths in about 10% of them. So, approximately there are 1 million deaths per year due to tuberculosis [8, 9]. The Mycobacteria principally target the lungs, moreover, it has been observed that *M. tuberculosis* may also reach and affect other parts of the body, such as the kidney, spine, and brain. A few people get tuberculosis ailment long after getting contaminated, even before their immune system can battle against the TB bacteria. Others may get the ailment years after the fact when their immune system gets frail for some other reason [8, 10].

Tuberculosis possesses a genuine risk to human wellbeing and one of the main reasons for significant human demise on the planet. Moreover, the emergence of drug resistance and its relationship with HIV infections have intensified worldwide circumstances. Unfortunately, despite advanced modalities for diagnosis and treatment of TB, people are still suffering a lot. There are specific properties associated with MTB that has presented vast challenges to develop an efficient drug against Tuberculosis [11]. The major obstacles in TB treatment like screening of compounds with anti-tubercular activity, the long duration medication, the lack of predictive animal models, and insufficient information on the physico-synthetic properties required for successful bacterial penetration [12], are being encountered by the pharmaceutical scientist.

The danger of creating dynamic (active) Tuberculosis ascends to 30% in diabetes victims. Usually, 80% - 90% of the patient having an infection of drug-resistant tuberculosis are relieved by taking concentrated anti-toxin treatment [2]. However, treatment by antibiotics is dependent on a load of drug-resistant *M. tuberculosis* in the patient [13]. Therapy of anti-drug or multidrug-resistant Tuberculosis (MDRTB: impervious to isoniazid and rifampin) is increasingly perplexing and takes nearly 2 years of chemotherapy amalgamation [14, 15]. Thus, progressively viable medicines are necessary to avoid or the emergence of tuberculosis. Treatment of the significant levels of drug-resistant *Mycobacterium tuberculosis* contamination, which incorporate rifampin-resistant (Rif-TB), MDR-TB, and extensively resistant TB (XDR-TB) requires new medications method and approaches to combat [5, 10]. The development of new methods of treatment is a complex process as anti-tuberculosis drugs are mostly given in combination to inhibit the further emergence of drug-resistant TB [14]. Moreover, dormant TB has also been observed in many people in which TB is not in a dynamic position and do not show any symptoms in a patient [16] whereas dynamic TB happens when the body cannot possess the

TB pathogen but at this condition, the bacteria can reproduce and cause wanted symptoms and people with dynamic TB can spread the contamination [9, 15]. In certain condition, some MTB strains are not affected by the treatment method and hard to treat tuberculosis [17, 18].

2. Need of research on new TB vaccine

In recent decades, advanced diagnosis and treatment method of TB has reduced the mortality rate up to significant level but TB still exists in world population causing extensive human suffering, economic burden led to global inequity. There are neonatal BCG vaccines that can prevent infants and young children from severe forms of TB but this vaccine is unable to show its effect in adolescents and adults who are crucial in TB transmission. We need to develop new efficient vaccines which could work in all age group people that may assist to fulfill the WHO end TB strategy that aims to reduce the TB mortality and TB incidences by 95% and 90% respectively worldwide.

Now, WHO is putting much efforts to produce TB vaccines and the Product Development for Vaccines Advisory Committee (PDVAC) is asking to develop a WHO preferred Product Characteristics (PPC) for new TB vaccines. The WHO's PPC data was established to document the crucial and priority requirements for vaccines which may show better safety and efficacy compared to BCG vaccine which is given to neonates and infants against pulmonary TB in adults, and new TB vaccines.

The major vaccine platforms like whole-cell vaccines, adjuvanted proteins, and recombinant subunit vector vaccines, are being considered in the pipeline of TB vaccine development. Now focus is on TB treatment in adolescents and adults by developing an effective candidate vaccine that may also replace the BCG in early life immunization. Many other aspects are in consideration in vaccine development, such as BCG boosters, reduction of treatment period using immunotherapeutic adjuncts and vaccine to prevent diseases reoccurrence in TB patient.

In recent developments, as per WHO report, there is TB vaccine candidate ($M72/AS01_E$) developed by the pharmaceutical company GlaxoSmithKline, in partnership with AERAS and was observed substantially effective against Tuberculosis disease and these results came out in a Phase IIb trial carried out in Kenya, South Africa and Zambia in patients having latent tuberculosis. This vaccine was found with 50% efficacy over about 3 years of continuous monitoring.

3. Globally situation of tuberculosis

According to the report of WHO, a sum of 1.4 million individuals passed on from TB in 2019 (counting 208,000 individuals with HIV). Around the world, TB is one of the top 10 reasons for death and the main source from a solitary irresistible specialist (above HIV/AIDS). In 2019, an expected 10 million individuals became sick with tuberculosis (TB) around the world. 5.6 million men, 3.2 million ladies and 1.2 million youngsters. In 2019, 1.2 million kids became sick with TB worldwide. The youngster and juvenile TB is frequently ignored by wellbeing suppliers and can be hard to analyze and treat. In 2019, the 30 high TB trouble nations represented 87% of new TB cases. Eight nations represent 66% of the aggregate, with India driving the tally, trailed by Indonesia, China, the Philippines, Pakistan, Nigeria, Bangladesh and South Africa. Multidrug-safe TB (MDR-TB) stays a general wellbeing emergency and a wellbeing security danger. A worldwide all out of 206 030 individuals with multidrug-or rifampicin-safe TB (MDR/RR-TB) were identified and told in 2019, a 10% expansion from 186 883 out of 2018. Internationally, the TB rate is falling at about 2% each year and somewhere in the range of 2015 and 2019, the combined decrease was 9%. This was not exactly most of the way to the End TB Strategy achievement of a 20% decrease somewhere in the range of 2015 and 2020. An expected 60 million lives were saved through TB analysis and treatment somewhere in the range of 2000 and 2019. Finishing the TB plague by 2030 is among the wellbeing focuses of the United Nations Sustainable Development Goals (SDGs). Tuberculosis generally influences grown-ups in their most gainful years. Nonetheless, all age bunches are in danger. More than 95% of cases and passings are in non-industrial nations. Multidrug-resistant tuberculosis (MDR-TB) is a type of TB brought about by microbes that do not react to isoniazid and rifampicin, the 2 best first-line hostile to TB drugs. MDR-TB is treatable and reparable by utilizing second-line drugs. Nonetheless, second-line treatment choices are restricted and require broad chemotherapy (as long as 2 years of treatment) with meds that are costly and poisonous.

Sometimes, more serious medication opposition can create. TB brought about by microbes that do not react to the best second-line hostile to TB medications can leave patients with no further treatment alternatives.

In 2019, MDR-TB stays a general wellbeing emergency and a wellbeing security danger. A worldwide total of 206 030 individuals with multidrug-or rifampicin-safe TB (MDR/RR-TB) were identified and advised in 2019, a 10% increment from 186 883 out of 2018. About portion of the worldwide weight of MDR-TB is in 3 nations – India, China and the Russian Federation.

Around the world, just 57% of MDR-TB patients are presently effectively treated. In 2020, WHO suggested another more limited (9–11 months) and completely oral routine for patients with MDB-TB. This exploration has shown that patients think that it's simpler to finish the routine, contrasted and the more drawn-out regimens that last as long as 20 months. Protection from fluoroquinolones ought to be rejected preceding the commencement of treatment with this routine.

As per WHO rules, the discovery of MDR/RR-TB requires the bacteriological affirmation of TB and testing for drug obstruction utilizing quick sub-atomic tests, culture strategies or sequencing advancements. Treatment requires a course of second-line drugs for at any rate 9 months and as long as 20 months, upheld by advising and checking for unfavorable occasions. WHO prescribes extended admittance to every single oral routine. Before the finish of 2019, 89 nations began utilizing more limited MDR-TB regimens and 109 had imported or begun utilizing bedaquiline, with an end goal to improve the viability of MDR-TB treatment.

4. The course of events in *Mycobacterium tuberculosis*

Mycobacterium tuberculosis basically passes through the 5 stages during its life cycle. At the first stage, the bacteria are inhaled through the air and typically engulfed by alveolar macrophages, further proceed to the symbiosis stage and causing the caseous necrosis in later stages. Eventually spread to other cells and causing rapid spread of diseases. The whole cycle is presented in detail in **Figure 1** and as a flow chart in **Figure 2**. The Mycobacterium gets entry into the lungs and resides in the alveoli of the lungs while it begins its primary infection. If the immune system fails to eliminate it then there are three cases observed with the mycobacterium in the alveoli. The first case could be the elimination phase, in which the immune system completely eliminates the infection. The next one retention phase where the immune system suppresses the infection but the bacteria remain viable and, in this case, the infection is known as Latent Tuberculosis which is the most asymptomatic Tuberculosis. And the third phase may involve Active infection, which makes the

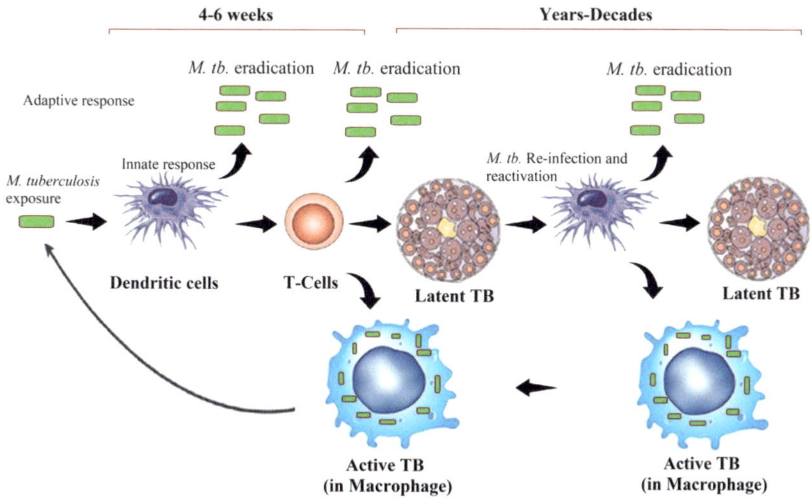

Figure 1.
Life cycle of Mycobacterium tuberculosis. This presentation is influenced with the figure available at online resource on study of the tuberculosis. (https://sites.google.com/site/mycobacteriumtbstudy/home/life-cycle-of-organism).

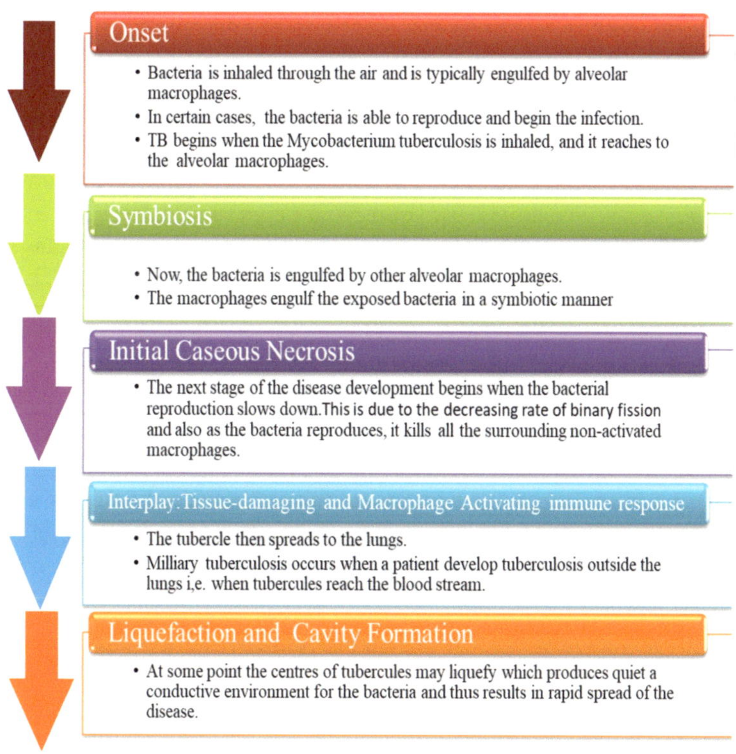

Figure 2.
Flow chart presentation of life cycle of Mycobacterium tuberculosis.

mycobacterium capable of evades the immune response and separates the infection in the lung tissue and at this point of active infection it is known as Active Tuberculosis [19–21].

M. tuberculosis has 5 stages in its life cycle as mentioned in **Figure 2** as flow chart [1, 2, 7, 22].

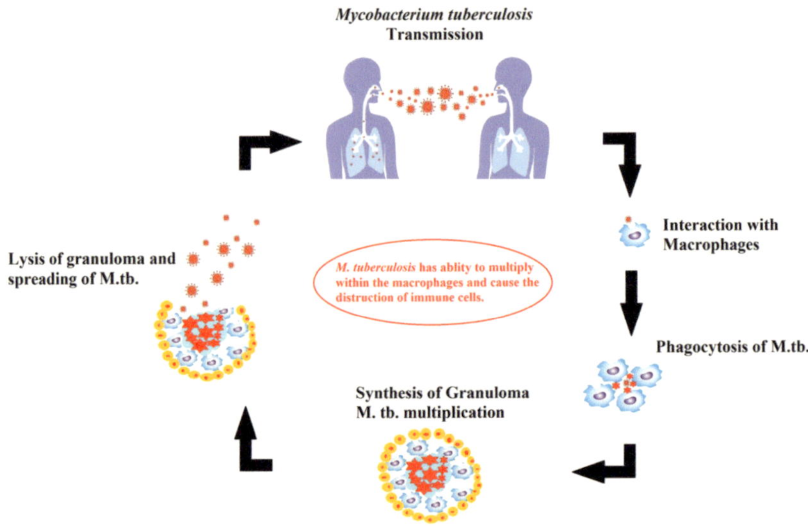

Figure 3.
Transmission of Mycobacterium tuberculosis. The representation is influenced with figure available in online resource. (https://www.istockphoto.com/in/vector/tuberculosis-life-cycle-of-mycobacterium-tuberculos-gm1200338165-343779875).

5. Pathogenesis and transmission of *Mycobacterium tuberculosis*

If somebody has active lung disease with TB they will cough and, in the cough, there would be infected droplets carrying the bacteria that could be inhaled by somebody else [8, 15]. Once the bacteria is inhaled it goes into the lungs and then it invades the normal mechanism for protecting lungs against bacterial infection which are the alveolar macrophages. It actively seeks out and invades these macrophages because it can prevent the normal macrophage killing mechanism. So, it diverts the normal figure lysosome pathways and that allows it to survive in the macrophage and it can be latent in that macrophage for decades [3]. Also, Macrophages because they move will allow the bacterium to spread bull RAC across the body and this is one of the reasons why sites of immune functions such as the lymph nodes often get infected with Tuberculosis and long-term persistence within the macrophages is led to latent diseases [18]. Besides, there is a certain inflammatory response to this infection which causes a very distinctive histologic appearance called granulomas and that is one of the hallmarks of Tuberculosis infection. Our closest infection is the presence of granulomas in the infected tissue [5, 6]. This transmission process is represented in **Figure 3**.

6. Mechanism of drug-resistant TB

This has been observed that various mechanism of drug resistance in M. tuberculosis is involved.

6.1 Presence of cell wall

The basic property leading to passive resistance to antibiotics in M. tuberculosis is because of its impervious cell wall [23]. The hydrophilic layer of arabinogalactan ensures the impervious nature of the cell wall to the surrounding hydrophobic substances. This layer is also present in hydrophobic mycolic acids which significantly prevents the entry of hydrophilic molecules [24]. This impervious nature of the cell wall results in the deposition of antibiotics throughout the cell wall, the accumulated antibiotics near the cell wall are removed steadily by the release of enzyme & with the involvement of several cellular components [25]. It is demonstrated that β-lactams, which act as inhibitors to the inclusion of peptidoglycan (responsible for maintaining the rigidity of the cell wall) into the cell wall, are degraded by the mycobacteria due to the presence of β-lactamases, which are the enzyme responsible for degradation of β-lactam antibiotics. Danilchanka et al. [24], reported the presence of CpnT channel protein in the outer membrane of both M. tuberculosis and *M. bovis*, which plays a dual role in nutrient absorption and selective sensitivity to antibacterial agents.

6.2 Slow metabolism mechanism

Bacteria that have long-generation time & undergo metabolic processes with a slower rate are estimated to be challenging targets for most of the antibiotics i.e., bacteria that are metabolically active and rapidly replicating act as a good target for antibiotics [26]. However, in M. tuberculosis, it is still unclear whether the long generation time confirms its resistance to drugs. However, it is been reported that the slow growth rate of M. tuberculosis plays a crucial role in drug resistance. For example, antibiotics such as carbapenems lose their activity comparatively at a faster rate than the growth rate of M. tuberculosis [27]. It is seen that certain specific genes which are involved in the production of triacyl-glycerol permit the growth of M. tuberculosis even in oxygen-deprived conditions. Triacylglycerol decline in the metabolic processes of M. tuberculosis.

6.3 Possession of numerous efflux pumps

These protein channels play a vital role in the regulation of normal metabolism and the physiology of the organism such as toxins, signaling molecules through the cell wall, residues, and nutrient transport [28]. Efflux pumps have shown adaptation to drug resistance in M. tuberculosis. Multi-drug efflux pumps serve as an outlet for cell antibiotics and usually pass through both the inner and outer membranes of the cell [29]. Regulatory protein systems are present in Drug-efflux proteins which are responsible for controlling the expression of the efflux pump and thus helps in specializing them for drug resistance roles [28].

6.4 Mutation in genetic materials

It has been shown that the acquisition of antibiotics resistance in M. tuberculosis is the result of spontaneous mutation in several chromosomal genes. This frequent mutation has been found to cause a deliberate alteration to the required interaction between each drug against tuberculosis and its specified target.

M. tuberculosis shows resistance to rifampicin due to mutation in rpoB of RNA polymerase, decelerating its affinity for rifampicin [30]. It has been identified in certain studies that specific codons can cause resistance to rifampicin only with the onset of mutation in them [31, 32]. Resistance to pyrazinamide is due to mutation in the pncA gene [33, 34]. The mutations in pncA gene account for the large number of resistance cases reported in Mycobacterial tuberculosis.

The mode of action of isoniazid resistance is complex and remains unclear, however, most strains of Mtb resistant to isoniazid are associated with a mutation in KatG and inhA [35, 36]. S315T of KatG mutation is more common in isoniazid-resistant strains. Mutation at this phase results in the formation of isoniazid product with a low affinity for isoniazid adduct [37].

Mutations in embB497 and embB406, codon 306 in embB and Polymorphism in embA, embC, are all involved in ethambutol resistance [38]. In 2013, Safi et al. proposed that the mutation in ubiA (Rv3806c) showed a high level of ethambutol resistance [39]. Some investigators have reported that the mutations in tlyA gene play a vital role in the resistance of Viomycin and Capreomycin [40, 41].

7. Extrapulmonary tuberculosis (EPTB)

TB as a rule influences the lungs, however, it can likewise influence different pieces of the body, like the brain, the kidneys, or the spine. An individual with TB can pass on if they do not get treatment. TB influencing any piece of the body other than lung parenchyma including different structure inside the chest like the pleura, pericardium and perihilar lymph hubs, alluded as extra aspiratory tuberculosis. EPTB incorporates tuberculosis meningitis, stomach tuberculosis (for the most part with ascites), skeletal tuberculosis, Pott's infection (spine), scrofula (lymph-adenitis), and genitourinary (renal) tuberculosis. Scattered, or miliary tuberculosis regularly incorporates aspiratory and extrapulmonary locales. It is assessed that extrapulmonary tuberculosis (EPTB) represents 15–25% of all instances of TB. HIV patients, particularly with low CD4 tallies, have higher paces of EPTB. Youngsters are bound to have skeletal TB than grown-ups [42]. Approximately 10% of all TB cases have both pulmonary and extrapulmonary TB, and an additional 20% have EPTB without pulmonary involvement [2, 43].

8. Major limitations and considerations to work with *M. Tuberculosis*

Mycobacterium tuberculosis is a gradually developing bacteria which must be handled cautiously under exacting containment to minimize the hazard to research centre individual [4]. The bacterium can reproduce inside the macrophage and kill the immune cell. Another limitation presented by the bacteria in the innovative work of new drugs is the idea of its cell wall which is wealthy in lipids and ultimately makes the development of homogenous and single-cell culture and troublesome [2]. *M. tuberculosis* can evade the immune response and recreate inside macrophages coming about because of several bacterial variables which along these lines can modulate the immune reaction [4, 5]. Although *M. tuberculosis* is Gram-positive bacteria its cell wall resembles the external membrane of Gram-negative bacteria since it is composed of an asymmetric bi-layer containing particular mycolic acids, along with glyco-lipids, lipo-glycans, and proteins [3, 9]. Therefore, novel drugs with viability and quicker acting mechanism which can most likely work in the shorter-term and along these lines give better outcomes in the treatment are desperately required [7].

9. Possible opinion regarding the challenges of new drug discovery for tuberculosis

Besides, the development of XDR strains of M. tuberculosis, 5.4% of MDR-TB cases are discovered to be XDR-TB (World Health Organization, 2010, Ref. [3]). Multidrug and Extensive Drug-Resistant Tuberculosis: 2010 Global Report on Surveillance and Response (World Health Organization, 2010, Ref. [4]) is testing TB treatment programs in a few nations and even raises the chance of a re-visitation of a circumstance much the same as the pre-anti-microbial TB time [1]. As of now, MDR-TB is treated by a blend of eight to ten medications with treatments enduring up to 18 two years; just four of these medications were really evolved to treat TB5. Such imperfect treatment prompts practically 30% of MDR-TB patients to encounter treatment disappointment [44]. The treatment alternatives for XDR-TB are exceptionally restricted as XDR-TB bacilli are safe not exclusively to isoniazid and rifampicin, yet in addition to fluoroquinolones and injectables, for example, aminoglycosides. Furthermore, there are not kidding results with most MDR-TB and XDR-TB drugs, incorporating nephrotoxicity and ototoxicity with aminoglycosides, hepatotoxicity with ethionamide and dysglycaemia with gatifloxacin [45]. In this manner, the current circumstance requires the prompt distinguishing proof of new frameworks that can address arising opposition and furthermore requests the direct of suitable clinical preliminaries as verifiably not very many clinical examinations have been performed to assess the adequacy of medications in MDR-TB or XDR-TB patient gatherings. Improving the diagnostics with more extensive inclusion of medication vulnerability testing will likewise assist with tending to the high mortality of MDR/XDR-TB and control the development of obstruction.

Critical difficulties exist in TB drug revelation because of the idea of the causative bacterium. The absence of prescient models for compound section into mycobacteria is likewise a restricting variable since the direct trial proof is arduous to get. Creating essential guidelines around compound passage and efflux could help with improving hits from biochemical screens which need entire cell action, just as adjusting the synthetic properties needed for great pharmacokinetic properties [8].

10. Existing and upcoming tuberculosis drug regime

The present routine of medication for drug-sensitive Tuberculosis treatment was set up during the 1980s. This treatment process encompasses four levels of medications, isonicotinic acid hydrazide, rifampin, Ethambutol dihydrochloride and Pyrazinoic acid amide for six months of treatment (**Table 1**). The essential focus of Tuberculosis drugs is cell wall biogenesis, deoxyribonucleotide replication, ribonucleotide transcription, and protein synthesis [15, 46].

Treatment of drug-resistant or multidrug-resistant (MDR) tuberculosis is substantially further unpredictable [8]. The success of the treatment process relies upon the patient record and drug affectability. MDR-Tuberculosis needs therapy for a long time with a combination of 5 other medications. These second-line drugs will in general be progressively costly and incorporate Sirturo, 2-ethylthio-isonicotinamide, Seromycin, Moxifloxacino, and Streptomycine, just like cutting edge medications rifampin systemic and Myambutol [5, 46]. For MDR tuberculosis therapy, we need to go through at least 6 months long treatment process including various vaccinations. Some have been observed to show adverse effects like heart electrophysiology dysfunction and ototoxicity [10, 13].

11. Drug combination trials and standardization of TB regimens

The WHO-recommended formulations of anti-TB drugs and fixed-dose combinations (FDCs) of drugs appear in the *WHO Model List of Essential Medicines* (available at www.who.int/medicines/publications/essentialmedicines/en). The formulations and combinations of anti-TB drugs available in each country should conform to this list.

Drug	Drug property	Acting pH	Site of action
Isoniazid (H)	Bactericidal after 24 hrs with a high potency. Kills more than 90% of bacilli in first few days of treatment.	Both alkaline and acidic medium	Both intracellular and extracellular
Rifampicin (R)	Bactericidal within 1 hrs with high potency.	Both alkaline and acidic medium	Both intracellular and extracellular
Pyrazinamide (Z)	Bactericidal with a low potency	Acidic medium	Intracellular bacilli
Ethambutol (E)	Bacteriostatic with a low potency. Minimizes the emergence of drug resistance	Both alkaline and acidic medium	Both intracellular and extracellular
Streptomycin (S)	Bactericidal with a low potency	Alkaline medium	Extracellular bacilli

Table 1.
Current drugs and their property.

Normalized treatment implies that all patients in a characterized bunch get a similar treatment routine. Standard regimens have the accompanying benefits over the individualized solution of medications:

i. errors in remedy – and in this way the danger of advancement of medication opposition – are decreased;

ii. estimating drug needs, buying, circulation and checking are encouraged;

iii. staff preparing is encouraged;

iv. costs are decreased;

v. maintaining a regular drug supply when patients move to start with one region then onto the next is made simpler;

vi. outcome assessment is helpful and results are tantamount.

12. Pharmaco-kinetic and pharmaco-dynamic contemplations for tuberculosis medications

Pharmacokinetic (PK) and pharmacodynamics (PD) properties of a medicinal drug play a substantial role to propose its feasibility for medicinal purpose *In vivo* [47, 48]. Along with the PK/PD of any anti-tubercular drugs, medication also considers other factors like comorbid conditions, safety profile, oral bioavailability and metabolic strength [4, 10]. Oral administration is mostly preferred for advanced Tuberculosis medication whereas, oral bioavailability is critical to treat Tuberculosis [4, 46]. Solubility and gastrointestinal permeability are the two major factors that affect oral bioavailability. At present. Generally, the bioavailability of tablets Tuberculosis ranges from 40–90% and new drugs must show such property of bioavailability [2, 7]. The smaller successive dosing of drugs is suggested to improve the adhesion and recommend to have daily doses. An ideal TB medicine must transmit to the lungs, the site of the primary infection, and should have the ability to infiltrate the granuloma to reach, such as intracellular and extracellular bacilli in the centre of hypoxia and undoubtedly necrotic region [9]. Preferably, the adhesion of drugs compounds in the target tissue must be maintained at a chosen site at minimal inhibitory conditions [49, 50]. This approach is used to avoid the phenomenon of drug binding to plasma protein, inhibition of tissue diffusion and improving the half-life of medicine. Lipophilic drugs have a major portion in anti-tubercular drugs. PK/PD and mode of action determines the dose of drugs for the treatment [5, 6].

In terms of drug safety, an ideal drug for Tuberculosis should not show any acute toxicity or long duration for the treatment [47, 51]. Because of the global nature of Tuberculosis therapy, an excellent drug must not show drug–drug interaction with other chemically or biologically active TB drugs within the regime [7, 22].

13. Target identification

With the entire genomic sequence available for *Mycobacterium tuberculosis*, the potentiality to explore new targets for the development of antibiotic throughout the *M. tuberculosis* genome became convenient [9, 10]. Novel chemical entities & targets are expected to avoid resistance to existing drugs and therefore improve current treatments. An ideal target for the development of antibiotic must necessarily be in vivo, vulnerable to medicines and drug-effective [6].

Genetic screens trials are the preliminary step in manifesting which genetic products might be targeted at chemotherapy against tuberculosis. Howsoever, all the necessary genes are not equally vulnerable to pharmacological action [20]. Besides, the target should also be available for competitive or chemical inhibition. That is, the target must have the ability to bind with another molecule rather than its substrate [10, 52, 53]. The inhibition or initiation of the protein function with a possible concentration of the low molecular weight compounds results in cellular breakdowns, such as cell death leading to apoptosis or attenuated growth [14, 46]. Besides being susceptible to chemical inhibition, an anti-target screen inhibitor should also produce drug-like compounds with specificity to affect target function in the absence of interference with any host orthologs [5, 54].

14. Current status of tuberculosis drug discovery

Various strategies have been developed by researchers and investigators and they proposed combined drugs for clinical trials after screening. All these drugs have a specific mode of actions but at the same time, they also showed some side effect which is a challenging task for investigators (**Table 2**). Currently, about 7 new combinations of drugs are under clinical trials. These lead combinations have been recognized by several methods and differential screening [10]. Few screening methodologies are as follows:

14.1 SQ109

A combinatorial library entirely based on 1,2-ethylenediamines such as Ethambutol was examined on two high-throughput in-vitro analysis. The first evaluation involves dilution of bouillon to calculate minimal inhibitory concentration (MIC) contrary to *Mycobacterium tuberculosis* [55]. The subsequent measurement is based on iniBAC promoter, inhibition of cell wall and bioluminescent assays for high-throughput screening [56]. SQ109 was determined on this screen. But the

Mode of therapy	Implemented Drugs	Possible adverse effects
First line oral agents (Ref. 33, 34)	Isoniazid	Hepatotoxicity, dermatological, gastrointestinal, hypersensitivity
	Rifampicin	Heartburn, epigastric distress, Thrombocytopenia, Leukopenia, hemolytic anemia, Menstrual disturbances etc.
	Ethambutol	Retrobulbar neuritis, gastrointestinal disturbance.
	Pyrazinamide	GI disturbances, Thrombocytopenia, sideroblastic anemia, Mild arthralgia myalgia etc.
Injectable anti-TB drugs	Streptomycin	burning, crawling, itching, numbness, prickling etc.
	Kanamicin	pain or irritation
	Amikacin	diarrhea, hearing loss, spinning sensation (vertigo), numbness etc.
Fluroquinol drugs	Ofloxacin	Nausea, diarrhea, constipation, gas, vomiting etc.
	Levofloxacin	low blood sugar, headache, hunger, sweating, irritability etc.
	Gatifloxacin	red, irritated, itchy, or teary eyes, blurred vision, eye pain etc.
Second line oral drugs	Ethionamide	Nausea, vomiting, diarrhea, abdominal/stomach pain etc.
	Prothionamide	depression and hallucinations
	Cycloserine	Headache, drowsiness, dizziness, or shaking etc.
	p-Aminosalicylic acid	persistent nausea, vomiting and diarrhea etc.
Anti-TB drugs with long term safety	Linezolid	severe diarrhea or diarrhea that is watery or bloody, fungal infections, low platelet counts etc.
	Redaquiline	Nausea / Vomiting, Dizziness, Headache, Hemoptysis etc.
	Clofazimine	diarrhea, nausea, vomiting, gastrointestinal intolerance etc.
	Amoxicillin	severe skin rash, itching, hives, difficulty breathing or swallowing etc.
	High dose Isoniazid	increased blood levels of liver enzymes and numbness etc.

Table 2.
Current mode of therapy and therapeutic drugs for tuberculosis.

mode of action and efficacy of SQ109 differ widely from ethambutol [57, 58]. SQ109 is bactericidal in nature and works by targeting a transmembrane transport protein MmpL3 which is responsible for transmitting trehalose monomicolate during cell wall synthesis [59, 60]. It acts against extracellular as well as intracellular bacilli and works on acute and chronic mouse models of tuberculosis infection [61]. SQ109 improved the pharmacological efficacy of the present four available first-line drugs against tuberculosis and represents synergy with Sirturo. It is presently under phase 2 clinical studies [5, 15].

14.2 Q203

It is an amide compound of imidazopyridine and was recognized by the whole-cell screening of infected macrophages [17]. Q203 prevents ATP synthesis via causing an interruption in the electron transport chain and thus also inhibits the cytochrome bc1 complex involved in the electron transport mechanism. Q203 possess an exceptional Pharmacokinetic profile and prevents bacterial replication [2, 20].

14.3 TBA-7371

A member of a series of 1,4-azaindole which was recognized by a strategy of transformation of scaffolds preceded by a program of optimization of lead of a compound imidazopyridine [62]. TBA-7371 inhibits DprE1 non-covalently, a decaprenyl phosphoryl-β-Dribose2′-epimerase, in cell wall Arabian biosynthetic pathway. TBA-7371 is bactericidal and is working against both acute and chronic mouse models of tuberculosis infection. It is under phase 1 clinical studies [3, 46, 57].

14.4 OPC-167832

It is a derivative of 3,4-dihydrocarostyril. OPC-167832 is bactericidal and works by targeting DprE1, leading to the prevention of mycobacterial infection. It represents improved performance when in combination with delamide. Presently it falls under the category of phase 1 clinical studies [10, 15].

14.5 GSK-070

It targets leucyl-tRNA synthetase and is an oxaborole derivative. Oxaborols block leucyl-t-RNA synthesis and ultimately results in blocking protein synthesis by constructing an adduct with t-RNA. It is active against both acute and chronic tuberculosis infection [10, 63].

14.6 PBTZ-169 & BTZ-043

They belong to benzothiazinones and were diagnosed from a broth dilution evaluation in vitro for the detection of antibacterial and antifungal activities. Benzothiazinones basically prevents the formation of arabinose involved in the biosynthesis of cell wall by covalently targeting DprE1. Both PBTZ-169 and BTZ-043 are bactericidal thus prevents bacterial replication and multidrug-resistant tuberculosis infection. They represent almost equal potency against isoniazid and rifampin in the mouse models of recurrent tuberculosis infection. PBTZ-169 is under phase 1 scientific studies [7, 9, 63].

15. Risk factors associated with tuberculosis treatments

Recent emigration makes Tuberculosis very likely to reactivate. Vitamin D deficiency has the same effect because vitamin D is an immune modulator and deficiency of that weakens the immune system, thus protecting against tuberculosis [3, 9]. Another factor, HIV infection, which is present in 8% of patient cases of tuberculosis and this problem of HIV allowing TB to be reactive and become a problem is actually before the patient has become heavily immune-suppressed [64]. Smoking, diabetes and the elderly are all examples where the immune system has been weakened to a degree and allows the potential infection to take hold and cause a problem [22, 63]. Homelessness drug abuse, alcoholism and other immune suppression steroids after transplantation to mention corrosive tumor necrosis factor treatment, all make an individual more likely to reactivate latent disease, like tuberculosis [6]. The antibiotics being used for the TB treatment have also shown some of the side effects and the present major challenge to researchers to overcome these drawbacks of antibiotics (**Table 2**).

16. Recent developments in diagnostic approaches for tuberculosis

It is not easy to conduct a clinal diagnosis of tuberculosis very frequently as confirmed diagnosis requires culturing the bacteria *M. tuberculosis* in a sample from the patient [5] and which is very slow-growing. For lung diseases, we take morning sputum for culture purpose and microscopic studies. We also have to do Biopsies of the affected tissues, because that will provide us with a sample for culture and also for looking histologically for the characteristic presence of granulomas [53]. Mycobacterial culture confirms the presence of mycobacterium in given samples by microscopy analysis, and we may also draw the resistance profile i.e., whether the present strain belongs to the sensitive *M. tuberculosis* group or has resistance to some drugs that can be used to treat it. The major obstacles in culture MTB are that it is slow-growing bacteria and may take 3–4 weeks in liquid culture media [51]. The acid-fast bacilli of mycobacterial infection are detected by the microscopy analysis whereas latent tuberculosis disease is identified by immunological responses to tuberculosis antigens, i.e., i) Heaf test / Montoux: cofounded by BCG ii) Interferon Gamma Release Assays (IGRA) [48, 49]. There are some tools developed recently for the detection of drug-resistant MTB that facilitates early detection too, such GeneXpert, line prob. assay, LAMP assay etc.

16.1 GeneXpert

GeneXpert can detect mutations that cause resistance against Rifampicin. The test is a molecular TB test that detects the DNA of *Mycobacterium tuberculosis*. It uses a sputum sample and thus provides result in less than 2 hours. It can also detect the genetic mutations which are associated with drug Rifampicin resistance [65]. WHO recommended that this test should be used as the primitive diagnosis test in individuals suspected of having Multi-drug resistant TB, or HIV associated TB.

16.2 Line probe assay (LPA)

This technique also helps to detects mutations causing resistance against Rifampicin. Moreover, this assay can also detect mutations related to drug isoniazid [66]. The line probe assay (LPA), is typically based on strip technology and thus is used in the diagnosis of TB. It also detects RIF as well as Isoniazid (INH) resistance caused due to mutations in *rpo*β, and both *inhA* and *katG* genes.

16.3 Loop-mediated isothermal amplification (LAMP) assay

WHO has recommended the TB-LAMP (loop-mediated isothermal amplification) test that requires minimal laboratory infrastructure and has been evaluated as an alternative to sputum smear microscopy, which remains the most widespread test used in resource-limited environments. TB-LAMP is a unique temperature-independent way to amplify the DNA of tuberculosis patients. It is a manual test that takes less than one hour and results can be visualized with the naked eye under UV light. The potent TB-LAMP instrument can be used at the level of the peripheral health center where microscopy is often performed. (https://www.who.int/tb/features_archive/TB_LAMP/en/).

17. Available treatment

At present treatment of tuberculosis requires more than one antibiotic with prolonged combination therapies to eradicate the infection and prevent resistance [58] and the standard therapy may include 4 antibiotics i.e., Isoniazid & Rifampicin (most effective drugs and these are given for six months and thus these two helps in killing the bacteria), Pyrazinamide & Ethambutol (given for first two months only) [67, 68]. During treatment antibiotics are required for a long period, the minimum treatment period is six months and if the patient is having CNS or bone disease it often goes on for at least 12 months [69]. The patient is asked to take four drugs for two months and then followed by two drugs for four months, and the actual dose given to the patient is decided by their body weight such as if a patient is lower than 50 kg, they get a lower dose while if the patient is above 50 kg, they get a higher dose [13].

Corticosteroids are given to patients with CNS or pericardial disease because this reduces the further chances of having long term brain damage. All the cases need to be monitored and notified so that there can be a screening process of the patient's close contacts as well [14, 17].

18. Conclusion

M. tuberculosis is a difficult pathogen to combat and the frontline drugs currently in use are between 40 and 60 years of age. There is an urgent need of novel tuberculosis drugs, but the time to identify, develop and ultimately advance new drug regimens on the market has been extremely slow in the past decade. Organic biochemistry remains to be performed to know the mechanism of activity, to empower lead advancement, and to ensure in vivo effectiveness [20]. Current efforts to develop drugs against tuberculosis are not enough to end the global tuberculosis epidemic. Due to the diversification and complexity of the infection for *M tuberculosis*, no model can completely define the in-vivo conditions in which mycobacteria are found in Tuberculosis patients and there is no sole standard detection condition for generating successful compounds for tuberculosis drug development. Recent efforts have focused on the development of whole-cell screening trials because objective-based biochemical screens of inhibitors over the past two decades have not provided new tuberculosis drugs [68]. There are significant challenges in the discovery of anti-tuberculosis drugs due to the nature of the causative bacteria. The lack of predictive models for the entry of compounds into mycobacteria is also a limiting factor. Several additional barriers in the development of tuberculosis drugs include: there are no well-established (PK)– (PD) paradigms, lack of validation and human-like pathology of animal models currently available for drug discovery, lack of clinical laboratories suitable for clinical trials, and the lack of adequate research funds. The biggest challenge in the development of anti-tuberculosis drugs is to reduce the duration of treatment for patients with drug-sensitive tuberculosis [18]. Noval drugs are needed to achieve this and overcome drug resistance. In addition, it should be possible to use new drugs for patients with HIV/AIDS co-infection. The present condition of tuberculosis drug development is far better than what was seen past 10–15 years ago. Howsoever, the development is still lacking behind because of the significant challenges in the drug discovery against drug-resistant tuberculosis and the shorter duration of the treatment required for tuberculosis prevention [12, 13].

We need to identify essential Tuberculosis targets based on better knowledge of the disease pathogen and physiology, develop sharp screening trials, and prepare compounds specifically designed to provide better clues for antibacterial activities [11].

Recent granuloma models are based on a single cell type to imitate the aggregate complex that is formed. Biomedical engineering methods can produce further diversified but still organized multicellular structures that clearly defines the organization of human granulomas. The challenge is that the need is urgent, but the process of discovery and development requires an excessive number of resources and time. The search for more effective vaccines should continue to provide long-term solutions to tuberculosis. At the same time, the development of drugs and regimes must be accelerated with a clearer approach [1, 9].

Acknowledgements

Dr. Manish Dwivedi thanks to DST-INSPIRE Faculty award.

Conflict of interest

The authors declare no conflict of interest.

Author details

Manish Dwivedi* and Priya Giri
Amity Institute of Biotechnology, Amity University Uttar Pradesh, Lucknow, India

*Address all correspondence to: mdwivedi@lko.amity.edu

References

[1] Koul A, Arnoult E, Lounis N, et al. The challenge of new drug discovery for tuberculosis. Nature. 2011 Jan 27;469(7331):483-490. DOI:10.1038/nature09657

[2] CDC "Tuberculosis Fact Sheets", Centers for Disease Control and Prevention, 2014; https://www.cdc.gov/tb/publications/ factsheets/general/ltbiandactivetb.htm accessed: 22 December 2017.

[3] World Health Organization. Multidrug and Extensively Drug-resistant TB (M/XDR-TB): 2010 Global Report on Surveillance and Response. WHO/HTM/TB/2010.3. Geneva, Switzerland: WHO; 2010.

[4] WHO "Multidrug-Resistant Tuberculosis (MDR-TB) 2016 Update", World Health Organization, 2016; http://www.who.int/tb/ challenges/mdr/mdr_tb_factsheet.pdf accessed: 29 September 2017.

[5] Sarathy, J. P.; Zuccotto, F.; Hsinpin, H.; Sandberg, L.; Via, L. E.; Marriner, G. A.; Masquelin, T.; Wyatt, P.; Ray, P.; Dartois, V. Prediction of Drug Penetration in Tuberculosis Lesions. ACS Infect. Dis. 2016, 12;2(8):552-63.

[6] Harvey AL, Edrada-Ebel R, Quinn RJ. The re-emergence of natural products for drug discovery in the genomics era. Nat Rev Drug Discov. 2015 Feb;14(2):111-129.

[7] Pai, M.; Behr, M. A.; Dowdy, D.; Dheda, K.; Divangahi, M.; Boehme, C. C.; Ginsberg, A.; Swaminathan, S.; Spigelman, M.; Getahun, H.; et al. Tuberculosis. Nat. Rev. Dis. Primers 2016, 2, 16076.

[8] Kumar, A., Chettiar, S., & Parish, T. (2016). Current challenges in drug discovery for tuberculosis. Expert Opinion on Drug Discovery, *12*(1), 1-4.

[9] Early JV, Casey A, Martinez-Grau MA, et al. Oxadiazoles have butyrate-specific conditional activity against *Mycobacterium tuberculosis*. Antimicrob Agents Chemother. 2016Jun;60(6):3608-3616.

[10] Yuan, T., & Sampson, N. S. (2018). Hit generation in TB drug discovery: From genome to granuloma. Chemical Reviews, *118*(4), 1887-1916.

[11] Rybniker, J.; Chen, J. M.; Sala, C.; Hartkoorn, R. C.; Vocat, A.; Benjak, A.; Boy-Rottger, S.; Zhang, M.; Szekely, R.; Greff, Z.; et al. Anticytolytic screen identifies inhibitors of mycobacterial virulence protein secretion. Cell Host Microbe 2014, 16, 538–548.

[12] Abrahams GL, Kumar A, Savvi S, et al. Pathway-selective sensitization of *Mycobacterium tuberculosis* for target-based whole-cell screening. Chem Biol. 2012 Jul 27;19(7):844-854.

[13] Mdluli K, Kaneko T, Upton A. Tuberculosis drug discovery and emerging targets. Ann N Y Acad Sci. 2014 Sep;1323:56-75.

[14] Tonge, P. J. Drug–target kinetics in drug discovery. ACS Chem. Neurosci. 2018

[15] Sala C, Dhar N, Hartkoorn RC, et al. Simple model for testing drugs against nonreplicating *Mycobacterium tuberculosis*. Antimicrob Agents Chemother. 2010 Oct;54(10):4150-4158.

[16] Gold, B.; Nathan, C. Targeting phenotypically tolerant Mycobacterium tuberculosis. Microbiol. Spectr. 2017, 5, 0031–2016.

[17] Rao SP, Lakshminarayana SB, Kondreddi RR, Herve M, Camacho LR, Bifani P, Kalapala SK, Jiricek J, Ma NL, Tan BH, Ng SH, Nanjundappa M, Ravindran S, Seah PG, Thayalan P,

Lim SH, Lee BH, Goh A, Barnes WS, Chen Z, Gagaring K, Chatterjee AK, Pethe K, Kuhen K, Walker J, Feng G, Babu S, Zhang L, Blasco F, Beer D, Weaver M, Dartois V, Glynne R, Dick T, Smith PW, Diagana TT, Manjunatha UH. Indolcarboxamide is a preclinical candidate for treating multidrug-resistant tuberculosis. Sci Transl Med. 2013 Dec 4;5(214):214ra168.

[18] Sledz, P.; Silvestre, H. L.; Hung, A. W.; Ciulli, A.; Blundell, T. L.; Abell, C. Optimization of the Interligand Overhauser Effect for Fragment Linking: Application to Inhibitor Discovery against *Mycobacterium tuberculosis* Pantothenate Synthetase. J. Am. Chem. Soc. 2010, 7;132(13):4544-5.

[19] Cooper AM. Cell-mediated immune responses in tuberculosis. Annu. Rev. Immunol. 2009;27, 393-422.

[20] Cragg GM, Newman DJ. Natural products: A continuing source of novel drug leads. BiochimBiophys Acta. 2013 Jun;1830(6):3670-3695.

[21] Protopopova, M.; Hanrahan, C.; Nikonenko, B.; Samala, R.; Chen, P.; Gearhart, J.; Einck, L.; Nacy, C. A. Identification of a new Antitubercular drug candidate, SQ109, from a combinatorial library of 1, 2-Ethylenediamines. J. Antimicrob. Chemother. 2005;56(5):968-974.

[22] Hoagland, D. T.; Liu, J.; Lee, R. B.; Lee, R. E. New Agents for the Treatment of Drug-Resistant *Mycobacterium tuberculosis*. Adv. Drug Delivery Rev. 2016;1;102:55-72.

[23] Maitra A, Munshi T, Healy J, Martin LT, Vollmer W, et al. (2019) Cell wall peptidoglycan in Mycobacterium tuberculosis: An Achilles' heel for the TB-causing pathogen. FEMS Microbiol Rev 43(5): 548-575.

[24] Danilchanka O, Pires D, Anes E, Niederweis (2015) The *Mycobacterium tuberculosis* Outer Membrane Channel Protein CpnT Confers Susceptibility to Toxic Molecules. Antimicrob Agents Chemother 59(4): 2328-2336.

[25] Chiaradia L, Lefebvre J, Marcoux J, Burlet Schiltz O, Etienne G, et al. (2017) Dissecting the mycobacterial cell envelope and defining the composition of the native mycomembrane. Scientific Reports 7(1): 12807.

[26] Pandey AK, Sassetti C M (2008) Mycobacterial persistence requires the utilization of host cholesterol. Proc Natl Acad Sci USA 105(11): 4376- 4380.

[27] Baek SH, Li A H, Sassetti CM (2011) Metabolic regulation of mycobacterial growth and antibiotic sensitivity. PLoS Biol 9(5): e1001065.

[28] Nguyen L (2016) Antibiotic resistance mechanisms in M. tuberculosis: An update. Arch Toxicol 90(7): 1585-1604.

[29] Knutson KL, Hmama Z, Herrera Velit P, Rochford R, Reiner NE (2014) Lipoarabinomannan of Mycobacterium tuberculosis promotes protein tyrosine dephosphorylation and inhibition of mitogen-activated protein kinase in human mononuclear phagocytes: Role of the Src homology 2 containing tyrosine phosphatase 1. Journal of Biological Chemistry 273(1): 645-652.

[30] Damtie D, Woldeyohannes D, Mathewos B (2014) Review on molecular mechanism of first line antibiotic resistance in Mycobacterium tuberculosis. Mycobact Dis 4(6): 174.

[31] Somoskovi A, Parsons LM, Salfinger M (2001) The molecular basis of resistance to isoniazid, rifampin, and pyrazinamide in Mycobacterium tuberculosis. Respir Res 2(3): 164-168.

[32] Heep M, Rieger U, Beck D, Lehn N (2000) Mutations in the beginning of the rpoB gene can induce resistance to

rifamycins in both helicobacter pylori and Mycobacterium tuberculosis. Antimicrob Agents Chemother 44(4): 1075-1077.

[33] Tahir MK, Nayyer M, Sheed KA, Tanwir AM, Iqbal SM, et al. (2019) Pyrazinamide resistance and mutations in pncA among isolates of Mycobacterium tuberculosis from Khyber Pakhtunkhwa, Pakistan. BMC Infect Dis 19(1): 116.

[34] Huy NQ, Lucie C, Hoa T, Hung NV, Lan N, et al. (2017) Molecular analysis of pyrazinamide resistance in Mycobacterium tuberculosis in Vietnam highlights the high rate of pyrazinamide resistance-associated mutations in clinical isolates. Emerg Microbes Infect 6(10): e86.

[35] Hazbón MH, Brimacombe M, Bobadilla del Valle M, Cavatore M, Guerrero MI, et al. (2006) Population genetics study of isoniazid resistance mutations and evolution of multidrug-resistant Mycobacterium tuberculosis. Antimicrob Agents Chemother 50(8): 2640-2649.

[36] Seifert M, Catanzaro D, Catanzaro A, Rodwell TC (2015) Genetic mutations associated with isoniazid resistance in Mycobacterium tuberculosis: A systematic review. PloS one 10(3): e0119628.

[37] Bollela VR, Namburete EI, Feliciano CS, Macheque D, Harrison LH, et al. (2016) Detection of katG and inhA mutations to guide isoniazid and ethionamide use for drug-resistant tuberculosis. Int J Tuberc Lung Dis 20(8): 1099-1104.

[38] Plinke C, Cox HS, Zarkua N, Karimovich HA, Braker A, et al. (2010) embCAB sequence variation among ethambutol-resistant Mycobacterium tuberculosis isolates without embB306 mutation. J Antimicrob Chemother 65(7):1359-1367.

[39] Tulyaprawat O, Chaiprasert A, Chongtrakool P, Suwannakarn K, Ngamskulrungroj P (2019) Association of ubiA mutations and high-level of ethambutol resistance among Mycobacterium tuberculosis Thai clinical isolates. Tuberculosis 114: 42-46.

[40] Johansen SK, Maus CE, Plikaytis BB, Douthwaite S (2006) Capreomycin binds across the ribosomal subunit interface using tlyA -encoded 2'-O-methylations in 16S and 23S rRNAs. Molecular Cell 23: 173-182.

[41] Georghiou SB, Magana M, Garfein RS, Catanzaro DG, Catanzaro A, et al. (2012) Evaluation of genetic mutations associated with Mycobacterium tuberculosis resistance to amikacin, kanamycin and capreomycin: A systematic review. PLoS One 7(3): e33275.

[42] N. Jean Haulman, ... Charles M. Nolan, in The Travel and Tropical Medicine Manual (Fourth Edition), 2008, 391-406.

[43] Masahiro Narita, Christopher Spitters, in The Travel and Tropical Medicine Manual (Fifth Edition), 2017.

[44] Mitnick, C. et al. Community-based therapy for multidrug-resistant tuberculosis in Lima, Peru. N. Engl. J. Med. 348, 119-128 (2003)

[45] Ma, Z., Lienhardt, C., McIlleron, H., Nunn, A. J. & Wang, X. Global tuberculosis drug development pipeline: The need and the reality. Lancet **375**, 2100-2109 (2010).

[46] Payne DJ, Gwynn MN, Holmes DJ, et al. Drugs for bad bugs: Confronting the challenges of antibacterial discovery. Nat Rev Drug Discov.2007 Jan;6(1):29-40.

[47] Sacksteder, K. A.; Protopopova, M.; Barry, C. E.; Andries, K.; Nacy, C. A. Discovery and development of SQ109:

A new Antitubercular drug with a novel mechanism of action. Future Microbiol. 2012; 7(7):823-837.

[48] Ioerger, T. R.; O'Malley, T.; Liao, R.; Guinn, K. M.; Hickey, M. J.; Mohaideen, N.; Murphy, K. C.; Boshoff, H. I.; Mizrahi, V.; Rubin, E. J. Identification of new drug targets and resistance mechanisms in Mycobacterium tuberculosis. PLoS One 2013; 23;8(9):e75245.

[49] Tobin, D. M.; Ramakrishnan, L. Comparative pathogenesis of Mycobacterium marinum and Mycobacterium tuberculosis. Cell. Microbiol. 2008 ;10(5):1027-1039.

[50] 50Piton, J.; Foo, C. S.Y.; Cole, S. T. Structural studies of Mycobacterium tuberculosis DprE1 interacting with its inhibitors. Drug Discovery Today 2017;22(3):526-533.

[51] Pienaar, E.; Sarathy, J.; Prideaux, B.; Dietzold, J.; Dartois, V.; Kirschner, D. E.; Linderman, J. J. Comparing efficacies of moxifloxacin, levofloxacin and Gatifloxacin in tuberculosis granulomas using a multi-scale systems pharmacology approach. PLoS Comput. Biol. 2017; 17;13(8):e1005650.

[52] Angelo Iacobino, Giovanni Piccaro, Federico Giannoni, Alessandro Mustazzolu, Lanfranco Fattorini. *Mycobacterium tuberculosis* Is Selectively Killed by Rifampin and Rifapentine in Hypoxia at Neutral pH. *Antimicrobial Agents and Chemotherapy* 2017; 23;61(3):e02296-16.

[53] Raju Mukherjee, Anup Chandra Pal, Mousumi Banerjee. Enabling faster go/No-go decisions through secondary screens in anti-mycobacterial drug discovery. Tuberculosis 2017 ;106:44-52.

[54] Zheng, X.; Av-Gay, Y. New era of Tb drug discovery and its impact on disease management. Curr. Treat. Options Infect. Dis. 2016; 8,299-310.

[55] Moreira, W.; Ngan, G. J.; Low, J. L.; Poulsen, A.; Chia, B. C.; Ang, M. J.; Yap, A.; Fulwood, J.; Lakshmanan, U.; Lim, J. Target Mechanism-Based Whole-Cell Screening Identifies Bortezomib as an Inhibitor of Caseinolytic Protease in Mycobacteria. mBio 2015, 6, e00253−00215.

[56] Palmer, A. C.; Kishony, R. Opposing effects of target overexpression reveal drug mechanisms. Nat. Commun. 2014, 5, 4296.

[57] Sorrentino, F.; del Rio, R. G.; Zheng, X.; Matilla, J. P.; Gomez, P. T.; Hoyos, M. M.; Herran, M. E. P.; Losana, A. M.; Av-Gay, Y. Development of an intracellular screen for new compounds able to inhibit Mycobacterium tuberculosis growth in human macrophages. Antimicrob. Agents Chemother. 2015 Oct 26;60(1):640-645.

[58] Wang, F.; Sambandan, D.; Halder, R.; Wang, J.; Batt, S. M.; Weinrick, B.; Ahmad, I.; Yang, P.; Zhang, Y.; Kim, J. Identification of a small molecule with activity against drug-resistant and persistent tuberculosis. Proc. Natl. Acad. Sci. U. S. A. 2013, 110, E2510−E2517

[59] Hung, A. W.; Silvestre, H. L.; Wen, S.; Ciulli, A.; Blundell, T. L.; Abell, C. Application of fragment growing and fragment linking to the discovery of inhibitors of Mycobacterium tuberculosis Pantothenate Synthetase. Angew. Chem., Int. Ed. 2009, 48, 8452− 8456.

[60] Bonnett, S. A.; Ollinger, J.; Chandrasekera, S.; Florio, S.; O'Malley, T.; Files, M.; Jee, J.-A.; Ahn, J.; Casey, A.; Ovechkina, Y. A Target-Based Whole Cell Screen Approach to Identify Potential Inhibitors of *Mycobacterium tuberculosis* Signal Peptidase. ACS Infect. Dis. 2016; 9;2(12):893-902.

[61] Pavelka, M. S., Jr.; Chen, B.; Kelley, C. L.; Collins, F. M.; Jacobs, W. R., Jr

Vaccine efficacy of a lysine auxotroph of Mycobacterium tuberculosis. Infect. Immun. 2003; 71(7):4190-4192.

[62] Xie, Z.; Siddiqi, N.; Rubin, E. J. Differential antibiotic susceptibilities of starved Mycobacterium tuberculosis isolates. Antimicrob. Agents Chemother. 2005, 49, 4778–4780.

[63] Cadena, A. M.; Fortune, S. M.; Flynn, J. L. Heterogeneity in tuberculosis. Nat. Rev. Immunol. 2017;17(11):691-702.

[64] Christophe T, Ewann F, Jeon HK, Cechetto J, Brodin P. High-content imaging of Mycobacterium tuberculosis-infected macrophages: An in vitro model for tuberculosis drug discovery. Future Med Chem. 2010 Aug;2(8):1283-1293.

[65] *Two hour detection of MTB* and *resistance* to *rifampicin*", *Cepheid International*, *2011* (www.cepheidinternational.com).

[66] Helb D, Jones M, Story E, Boehme C, Wallace E, Ho K, Kop J, Owens MR, Rodgers R, Banada P, Safi H, Blakemore R, Lan NT, Jones-López EC, Levi M, Burday M, Ayakaka I, Mugerwa RD, McMillan B, Winn-Deen E, Christel L, Dailey P, Perkins MD, Persing DH, Alland D. Rapid detection of Mycobacterium tuberculosis and rifampin resistance by use of on-demand, near-patient technology. J Clin Microbiol. 2010 Jan;48(1):229-237.

[67] DeBarber, A. E.; Mdluli, K.; Bosman, M.; Bekker, L.-G.; Barry, C. E. Ethionamide Activation and Sensitivity in Multidrug-Resistant *Mycobacterium tuberculosis*. Proc. Natl. Acad. Sci. U. S. A. 2000; 15;97(17):9677-82.

[68] Tousif S, Ahmad S, Bhalla K, Moodley P, Das G (2015) Challenges of tuberculosis treatment with DOTS: An immune impairment perspective. J Cell Sci Ther 6: 223.

[69] Boshoff, H. I.; Myers, T. G.; Copp, B. R.; McNeil, M. R.; Wilson, M. A.; Barry, C. E., 3rd The transcriptional responses of Mycobacterium tuberculosis to inhibitors of metabolism: Novel Insights into Drug Mechanisms of Action. J. Biol. Chem. 2004, 279, 40174–40184.

Chapter 3

Study of Various Chemically and Structurally Diverse Currently Clinically Used and Recently Developed Antimycobacterial Drugs

Saad Alghamdi and Mohammad Asif

Abstract

Infectious diseases originate from pathogens and increased severely in current years. Despite numerous important advances in antimicrobial therapy, the extensive use and misuse of these antimicrobial drugs have caused the emergence of microbial resistance, which is a serious risk to public health. In particular, the emergence of multidrug-resistant pathogens has become a serious difficulty in the therapy of pathogenic diseases. Therefore, the progress of novel drugs to deal with resistant pathogens has become one of the most essential areas of antimicrobial research today. In addition to the development of novel and efficient antimicrobial agents against multidrug-resistant pathogens, recent attention has focused on the treatment of tuberculosis. Therefore, recent developments have been directed towards examining currently used and newly developed antimycobacterial drugs and their toxicities and mechanism of action.

Keywords: Chemotherapy, epidemiology, tuberculosis, multidrug-resistant, Mycobacterium

1. Introduction

Tuberculosis (TB) is a chronic infectious and zoonotic disease caused by the *Mycobacterium tuberculosis* (*Mtb*) complex. It is responsible for a lung infection (pulmonary TB) and other body parts (extrapulmonary TB) [1, 2]. TB is the second cause of death next to human immunodeficiency virus (HIV) infection. TB is a global public health disaster since 1993 at a time of expected 7–8 million cases and 1.3–1.6 million deaths occurred yearly. In 2010, 8.8 million new cases of TB and 1.1 million deaths from TB, and 0.35 million deaths from that HIV-allied TB and worsen due to the maturity of anti-TB drug resistance (DR) [3]. The emergence of resistance against anti-TB drugs is a barrier to the success of TB treatment. Inadequate TB treatment is responsible for the incident of DR-TB [4]. According to the World health organization (WHO) report, TB has spread to every area of the world. As much as one-third of the world's population is presently infected, more than any other infectious microbial disease [5]. These numbers however are only a partial representation of the world TB threat. It was predictable that nearly 1 billion more people will be infected with TB in the coming 20 years. However, the number of new TB cases is still growing slowly, 95% occur in developing nations every year, and about one million young women per year are offended with this infectious disease in the developing world. The occurrence of TB is associated with a dense population, deprived sanitation, and, poor diet [6]. Directly observed treatment, short-course (DOTS), is the effective way of the control of TB. The three main anti-TB drugs, isoniazid (INH), rifampicin (RIF), pyrazinamide (PZA) are used. These drugs are hepatotoxic and may cause drug-related hepatitis. Despite the success of the DOTS therapy, the appearance of MDR-TB strains, persistently isolated from the sputum of pateints, darkens the future [7]. The expansion in TB incidence during current years is primarily due to the incidence of TB in synergy with the HIV pandemic, which increases the risk of growing the new TB cases were attributable to HIV co-infection, and as well as the emergence of MDR-TB strains [8, 9]. Therefore, the objective of this paper is to review the current status of antimycobacterial drugs.

2. Etiology and routes of transmission

Tuberculosis (TB) is caused by the *Mtb* complex. The mycobacterium is non-motile, Gram-positive, rod-shaped, obligate aerobic bacteria that belong to the order Actinomycetales and family Mycobacteriaceae. This *Mtb* complex includes (subspecies *M. canetti*), *M. bovis*, *M. microti*, *M. africanum*, *M. caprae*, *M. bovis* BCG, and *M. pinnipedii* [10]. The cough in a TB patient is caused by the infection of *Mtb* and distributed to air during coughing. The healthy persons who inhaled air droplets of TB infected person and make contact become infected [11].

Tuberculosis-HIV Combination:

The current opinion disclosed that one-third of the 42 million people living with HIV/AIDS all around the world are co-infected with TB. As per the WHO report, about 90% of the patients containing both TB and HIV died within only some months after clinical indications have arisen. Thus, WHO warned the world of the "even bigger TB-HIV crisis" and explained for extensive accessibility of free anti-TB drugs to individuals living with HIV. The HIV cases are spreading quickly in India with the biggest number of TB cases all around the world [12–15].

3. Chemotherapy of tuberculosis

First-line anti-TB drugs:

Treatment of TB is mainly dependent on first-line anti-TB drugs (**Figure 1**), which comprises SM, INH, RMP, EMB, and PZA, these are more effective and less toxic effects as compare to second-line anti-TB drugs [15].

Second-line anti-TB drugs:

According to the WHO, there are six drugs of second-line anti-TB drugs. These drugs are categorized as second-line anti-TB drugs due to one of two potential reasons: 1) they are less active than the first-line anti-TB drugs or more toxic side-effects or 2). These drugs involve different classes namely, aminoglycosides (**Figure 1**): (kanamycin, amikacin), polypeptide analogs (**Figure 2**): (viomycin, capreomycin), FQs (**Figure 3**): (CPX, MXF, OFX, etc), thioamides: (prothioamide, ethionamides), cycloserine and para-aminosalicylic acid (**Figure 4**) [2–4].

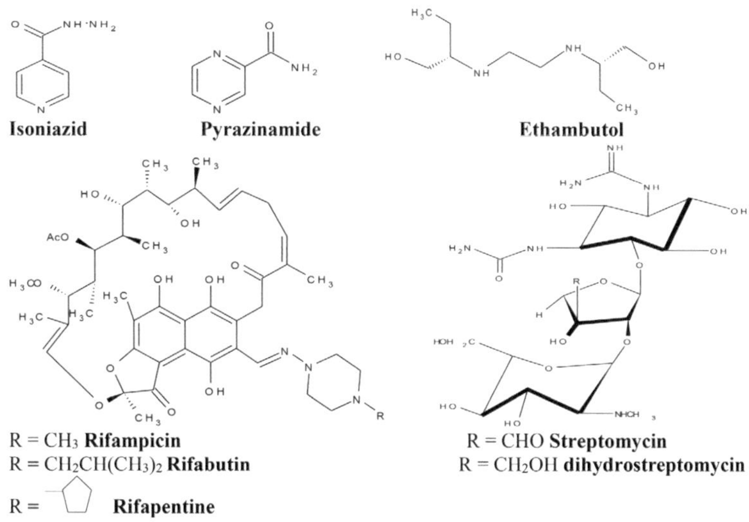

Figure 1.
Structures of first-line anti-TB drugs.

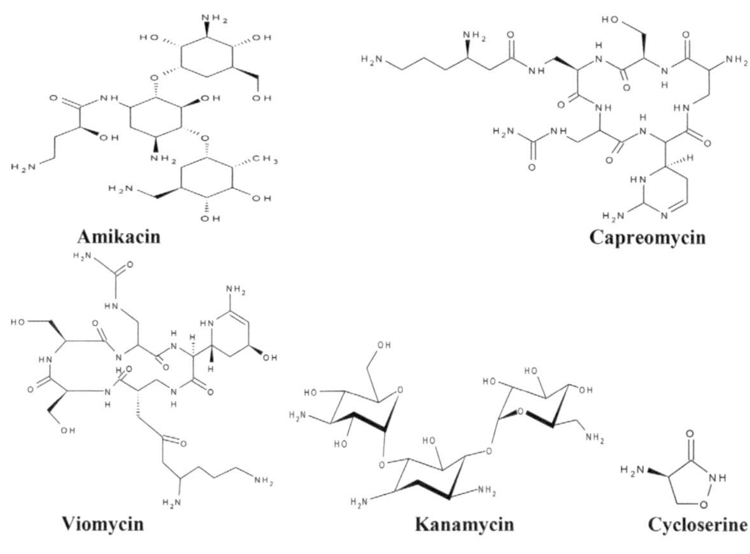

Figure 2.
Structures of different antibiotics (second-line drugs).

Drugs for HIV/TB

Clarithromycin (**Figure 5**) is a macrolide antibiotic drug used in HIV infected TB patients to cure the *M. avium* complex (MAC). It has an analogous of erythromycin but is more efficient against certain Gram-negative bacteria, mainly *Legionella pneumophila*. Thioacetazone (**Figure 5**) is valuable in stopping resistance to more influential drugs like INH and RIF. It is not at all used on its own to treat TB. It uses is declining because it can originate severe skin reactions in HIV positive patients. It is also identified to kill MDR-TB. It is no longer suggested for treatment due to its adverse effects like urination-difficulties, dry mouth, glaucoma, and postural hypotension. The circumstances are further complex by the emergence of MDR-TB and XDR-TB by infections with lethal synergy with HIV/AIDS [4, 15].

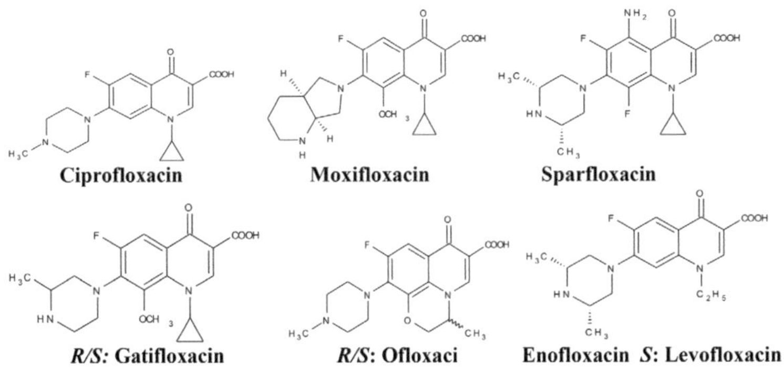

Figure 3.
Structures of different fluoroquinolones (second-line anti-TB drugs).

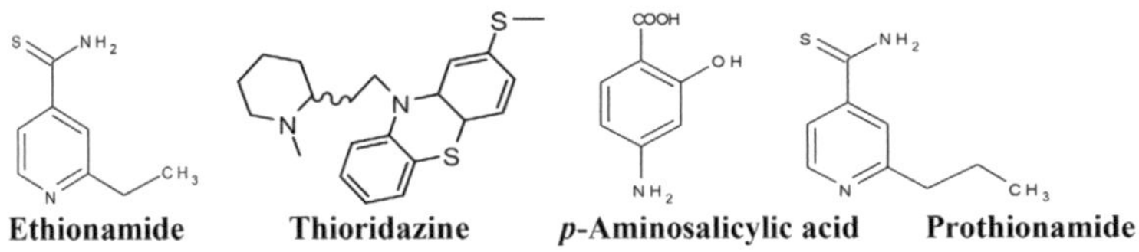

Figure 4.
Structures of different second-line anti-TB drugs.

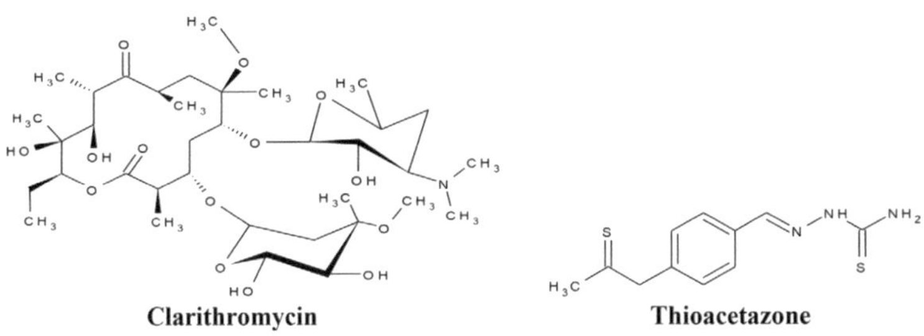

Figure 5.
Structures of drugs for HIV/TB.

4. Properties and mechanism of currently used common anti-TB drugs

4.1 Primary agents

Isoniazid (Nydrazid, Laniazid)

It is a bacteriostatic drug against resting cells and bactericidal against dividing microorganisms. Isoniazid (INH) is an anti-TB drug since 1952 and acts as a bactericidal and bacteriostatic for rapidly and slowly growing bacilli. It diffuses across the *Mtb* cell membrane [16]. The INH targets KatG and inhA gene. KatG gene encodes catalase/peroxidase enzyme that activates prodrug and peroxynitrite that are involved in pathways of reactive nitrogen and oxygen intermediates [17, 18]. InhA gene encodes NADH dependent enoyl-Acyl Carrier Protein (ACP)-reductase that causes inhibition of mycolic acid synthesis [19, 20]. The INH is a close to ideal antibiotic and very selective (MIC value ~0.025–0.05 μg/mL and other bacteria MIC vale >500 μg/mL). The INH has good oral availability and low toxicity. It Inhibits mycolic acid biosynthesis and targets the enoyl-acyl carrier protein reductase enzyme (InhA) engaged in mycolic acid synthesis. The INH inactivation of

IhhA needs metabolic activation. It is also utilized in combinations, INH and RIF and INH, RIF, and PZA.

Rifampicin

Rifampicin (RIF) was isolated from *Streptomyces mediterranei* from the soil sample and used as an anti-TB since 1972 [21]. It is still utilized as the best choice of anti-TB drugs. RIF diffuses across *Mtb* cell membrane due its lipophilic nature. RIF inhibits the mycobacterial transcription by binds to the β-subunit of DNA-dependent-RNA polymerase [22–24].

Rifamycins

Rifamycins are natural products from *Amicolaptosis mediterranei* belong ansamycin family and are. These are active towards various bacteria but used almost exclusively against TB [2–4].

Rifampin (Rifadin)

Rifampin is a semisynthetic analog of rifamycin and the most effective anti-TB agent with MIC values as low as 0.005 μg/mL. It is used as an oral or parenteral formulation, it can access CNS and it is sensitive to moisture [2–4].

Rifapentine (Priftin)

Rifapentine is a cyclopentyl analog of RIF. The benefit over rifampin is less repeated dosing. It inhibits bacterial DNA-dependent RNA polymerase and binds to the β subunit. The RIFs blocks the elongation of the RNA transcript and inhibits gene expression. It also acts as a CYP450 inducer. One remarkable side effect is the discoloration of body fluids. The RIFs are not suggested for use in HIV infected patients. Two RIF analogs are existing for indications other than TB [2–4].

a. **Rifabutin** (Mycobutin): It is used mostly in MAC infections [2–4].

b. **Rifaximin** (Xifaxan): Indicated for the treatment of traveler's diarrhea [2–4].

Ethambutol (Myambutol):

Ethambutol (EMB) is a bacteriostatic, active against growing bacilli, and used as an anti-TB drug in 1966. It obstructs polymerization of cell wall component lipoarabinomannan and arabinogalactan that interrupted biosynthesis of darabinofuranosyl-P-decaprenol and produced bacteriostatic effect [25, 26]. EMB (+) isomer is orally active, 16 times more potent than meso isomer, and 200 times more potent than (−) isomer. The EMB inhibits the polymerization of cell wall arabinan and results in the addition of the lipid carrier deca-prenol phosphoarabinose. The EMB interferes with the transfer of arabinose to the cell wall acceptors. EMB is effective only towards energetically dividing cells and its action is synergistic with RIF. Arabinosyl transferase enzyme is a target for the action of EMB in both *Mtbs* and *M. avium*. The enzyme is encoded by the embCAB gene organized as an operon and engaged in the arabinogalactan synthesis [27, 28].

Pyrazinamide (Aldinamide)

Pyrazinamide (PZA) is a pyrazine derivative of nicotinamide and its mechanism is assumed to be analogous to INH. It has to be metabolically activated and PZA-resistant strains of *Mtb* contain a mutation in the hydrolase gene. PZA activity against *Mtb* depends on the anaerobic and acidic conditions. PZA is activated to pyrazine acid by the pyrazinamide/nicotinamidase that encoded by gene pncA [29]. Acidic condition favors the production of protonated pyrazinoic acid that collected in the *Mtb* cell membrane which interrupts cell membrane potential and altered membrane transport [30]. The new target of PZA', clpC1 (Rv3596c) that encodes an ATP-dependent ATPase is liable for protein degradation by the complex structure with protease clpP1 and clpP2 [31, 32].

Streptomycin

Streptomycin (STR) is the first antibiotic cure for TB and it is isolated from the soil microbe *Streptomyces griseus* in 1943 [1]. It is active against growing bacilli, but not active against intracellular and non-growing bacilli. STR is still considered a first-line anti-TB drug but is used less frequently than the other drugs. It has no action against *M. avium*. Resistance is frequently due to phosphorylation. It targets both rpsl and rrs genes that encode 30S ribosomal protein S12 and 16S rRNA, respectively, and finally inhibit the instigation of the translation in the protein synthesis [33, 34].

Secondary/Retreatment Agents:

Aminosalicylic Acid (P.A.S. Parasal):

Para-aminosalicylic acid (PAS) is an oral drug that fell out of use because of adverse effects and frequent resistance. Related to sulfonamides, it is bacteriostatic and acts as a competitive inhibitor of mycobacterial dihydropteroate synthase. There are two mechanisms to produce the desired effect. First, it inhibited folic acid synthesis by the inhibiton of dihydrofolate synthase and dihydropteroate synthase that produces hydroxyl dihydrofolate antimetabolite responsible for the folic acid synthesis [35]. Secondly, it reduced the uptake of iron, that is essential for cell wall component mycobactin synthesis [15].

Ethionamide (Trecator SC):

Ethionamide (ETH) is developed as a derivative of INH but less potent than INH. Two genes play a role in the mechanism of actions ETH is ethA and inhA. EthA is regulated by the transcriptional repressor ETH [36]. The mechanism of action is like INH. The oxidative activation comes into sight it is by an enzyme other than KatG projected to form a covalent connection with InhA. The mechanism of action of the ETH is a disruption of mycolic acid synthesis by which monooxygenase enzyme activated ETH that binds to NAD+ and forms an adduct which inhibits enoyl acyl-ACP reductase enzyme [37–39] (**Figure 6**).

Cycloserine (Seromycin):

Cycloserine (CYS) is a natural compound and restricted to being retreatment because of CNS toxicity. CYS is a cyclic derivative of serine hydroxamic acid and terizidone. It is isolated from *Streptomyces orchidaceous* in the 1950s. Its acts by interrupting mycobacterial cell wall synthesis by inhibition of L-alanine racemase encoded by alrA that forms D-alanine from L-alanine and D-phenylalanine synthetase crucial for the production of peptidoglycan and cell wall synthesis by the inclusion of Dalanine into pentapeptide [40, 41]. It inhibits peptidoglycan formation, particularly-blocks the alteration of L-Ala to D-Ala.

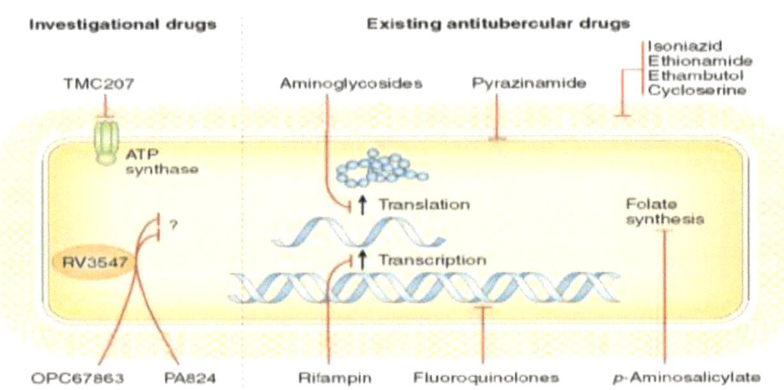

Figure 6.
Presently used anti-TB drugs and sites of action.

Fluoroquinolones (FQs) are the derivative of chloroquine in the 1960s and were used as bactericidal in human and veterinary medicines [42]. The FQs are acted by blocking of mycobacterial DNA replication by binding to α and β subunits of DNA gyrase (gyrA and gyrB), which catalyze the supercoiling of DNA and finally, inhibits DNA synthesis [43].

Aminoglycosides and polypeptides

The aminoglycosides (amikacin, kanamycin) and polypeptides (capreomycin, viomyocin) act by inhibiting protein synthesis. Kanamycin and amikacin alter 16S rRNA and capreomycin and viomycin interfere with small and large subunits of the 70S ribosome [44, 45].

Capreomycin (Capastat)

Capreomycin belongs to the tuberactinomycin family, a highly basic cyclic pentapeptide with a sixth amino acid side chain. It is the most active compound of this family and blocks protein synthesis and interferes with initiation tRNA selection and chain elongation. It binds to a site on 16S rRNA and the 23S rRNA. Some mycobacterium resistant to capreomycin is also resistant to kanamycin [2–4].

Kanamycin

The most commonly used second-line aminoglycoside and only given by intramuscular (IM) administration [14, 15].

Linezolid

Linezolid is an oxazolidinone derivative that interrupts the early stage in protein synthesis by binding to the 23S rRNA of the 50S subunit. The gene rplC and rrl are concerned in the action of Linezolid. The rplC gene that encodes 50S ribosomal L3 protein to involve in the synthesis of the ribosomal peptidyl-transferase. Hence, rrl gene has 3138 bp length that encodes 23S ribosomal RNA [46].

5. Compounds originating from existing families of drugs

Fluoroquinolones

Fluoroquinolones (FQs) were established into clinical applications in the 1980s and extensively used for the treatment of various bacterial infections [47]. The FQs have been also originated to have anti-TB activity [48] and are presently used as second-line anti-TB drugs. Cross-resistance has been accounted for within the FQs class such that reduced vulnerability to one FQ possibly presented reduced vulnerability to all FQ derivatives [49–51]. With the extensive use of FQSs for the therapy of common microbial infection, resistance to FQs remains uncommon and occurs mostly in MDR strains. The cross-resistance was observed among the various FQ compounds tested (OFX, LVX, GAT, MXF, and CPX) [52]. The rapid progress of resistance is mostly when FQs are used as the only active drugs in a failing multi-drug therapy [53–55]. These new agents are currently taken in concern as anti-TB drugs.

Gatifloxacin

Gatifloxacin (GAT) has bactericidal activity against *Mtb* [56]. It revealed the highest bactericidal effect during the first 2 days. GAT was used in combination with the first-line anti-TB drug INH or RIF: GAT was able to somewhat enhance the bactericidal activity of INH or RIF only for the first 2 days [57]. One study reported that when evaluated in mice in combination with ethionamide and PZA (high doses: 450 mg/kg, 5 days per week). The GAT was capable to clear the lungs of infected animals after 2 months of therapy [58].

Moxifloxacin

*In-vitro m*oxifloxacin (MXF) show to kill a subpopulation of tubercle bacilli that not killed by RIF, while the older FQs, (ciprofloxacin) CPX, and (ofloxacin) OFX did not have any major bactericidal effect. The MXF obstructs protein synthesis in gradually metabolizing bacteria through a mechanism that varies from that used by RIF. In mice models, the effect of MXF against tubercle bacilli was comparable to that of INH [59]. In combination with MXF and PZA has been killing the bacilli more successfully than the INH + RIF + PZA combination [60]. The substitution of MXF with INH in the standard drug therapy could relieve a probable antagonism among the presently used drugs [61]. The MXF might be a promising candidate drug to shorten TB treatment [62, 63].

Non-fluorinated quinolones

A series of 8-methoxy non-fluorinated quinolone analogs (NFQs) lack a 6-fluorine atom in their quinolone ring distinguishing them from fluorinated quinolone compounds such as GAT and MXF. The NFQs target a broad range of bacteria and they appear to operate preferentially through inhibition of DNA gyrase. The NFQs are presently being tested against *Mtb* [4].

Macrolides

The anti-TB effect of the macrolide antibiotics through the synthesis of additional chemically modified analog of erythromycin. Some analog were recognized as anti-TB agent superior to the clarithromycin [4, 15].

New rifamycin derivatives

Rifalazil, a semisynthetic RIF, is described by a long half-life and is more effective than RIF and rifabutin against *Mtb* strains [64]. However, high-intensity RIF–resistant strains present cross-resistance to all rifamycin drugs [65].

Bedaquiline or TMC207

Bedaquiline is a diarylquinoline and used bactericidal. Bedaquiniline involves blocking the proton pump of ATP synthase of *Mtb* then depletes the energy demand of both replicating and nonreplicating (dormant) mycobacteria and at the result in cell death [66, 67].

Delamanid or OPC 67683

Delamanid is a dihydro-nitroimidazooxazole derivative and activated by deazaflavin-dependent nitroreductase enzyme (Rv3547). It acts by interrupting the mycobacterial cell wall component synthesis. Delamanid inhibits the methoxy- and keto-mycolic acid synthesis which is a vital component of the *Mtb* cell wall. It is active against both growing and nongrowing mycobacteria [68, 69].

PA-824

PA-824 is a nitroimidazole derivative and it activated by deazaflavin-dependent nitroreductase like delamanid. Mechanism of action is not clearly konown but it could be described as its activity in replicating and non-replicating mycobacteria. In aerobically replicating cell PA-824 interrupts mycolic acid synthesis by the collecting of hydroxymycolates instead of ketomycolates [70, 71]. In hypoxic non replicating mycobacteria, PA-824 release nitric oxide (NO) that interferes with cytochrome oxidase to disrupt the energy metabolism of the cell wall [72].

SQ-109

SQ-109 is an ethambutol analog and its mechanism of action is not known. It has no inhibitory activition against the secreted Ag85 mycolyltransferase. Rather SQ-109 causes collection of trehalose monomycolate a precursor of trehalose dimycolate by obstructing accumulates of mycolic acids into the *Mtb* cell wall core [73].

Antitubercular drugs with the new and different moiety

To investigating useful drug candidate's currently in two major categories: Novel chemical entity and compounds instigating from existing relatives of currently used drugs, where novel chemistry is used to optimize the new compounds.

Nitroimidazopyrans and Nitroimidaoxazoles derivatives

In this series, the lead molecules are CGI 17341 and PA824/PA1343 and inhibit the cell wall synthesis. However, two key areas of concern also require to attend-possible mutagenicity resulting from the presence of a nitro group, and the chance for the development of drug resistance. The latter is encouraged by the reality that the nitroimidazoles induce a high rate of mutation [2–4], leading to uncertainties that this might cause the appearance of MDR bacteria. Since the drugs will certainly be used in combination therapy [74].

Nitroimidazole PA-824

The PA-824 is a nitroimidazole derivative and used as anti-TB agent. PA-824 acts mainly as synthesis of cell wall components inhibitor. *In vitro*, PA-824 showed high activity against drug-sensitive and MDR-TB strains, there is no cross-resistance with currently used anti-TB drugs. Moreover, PA-824 has shown *in-vitro* bactericidal activity against both replicating and static bacteria [4]. The PA-824 bactericidal effect against nonreplicating bacteria was equivalent to the RIF. Use of PA-824 as monotherapy in mice and cause reduced bacterial counts in the lungs is better than RIF or INH monotherapy. After 12 weeks of treatment with PA-824, RIF, or INH, complete removal of the bacterial load was not getting in any of the treated mice [2–4].

Nitroimidazoles CGI 17341

The CGI 17341 has substantial potential as anti-TB agent in a preclinical study. *In-vitro* at 0.04 to 0.3 μg/ml, it inhibited both drug-susceptible and MDR-TB strains and exhibited no cross-resistance with INH, RIF, SM, or EBM. *In-vitro* against *Mtb*, its action was similar to that of INH and RIF and higher to SM, norfloxacin, ciprofloxacin, and the oxazolidinone DuP 721. In *Mtb*-infected mice, oral treatment with CGI 17341 on days 11 and 12 after infection resulted in an ED_{50} of 7.7 mg/kg and showed a significant dose-dependent progress in survival time [75]. These drugs were not mutagenic and showed potent activity against replicating and static *Mtb*, including MDR strains.

Nitroimidazole OPC-67683

It is mycolic acid inhibitors and interferes with the biosynthesis of the mycobacterial cell wall. *In vitro*, OPC-67683 was exhibited high activity against drug-sensitive as well as DR strains with MICs varying 6–24 ng/mL. There is no cross-resistance with any of the current first-line anti-TB drugs. Moreover, OPC-67683 exhibited strong intracellular activity against the *Mtb*H37Rv residing within human macrophages and type II pneumocytes. The OPC-67683 is active against *Mtb*H37v and MDR-TB strains *in-vivo* starting from a concentration of 0.03125 mg/body [2–4]. The OPC-67683 exhibited 6–7 fold elevated activity compared to first-line drugs INH and RIF.

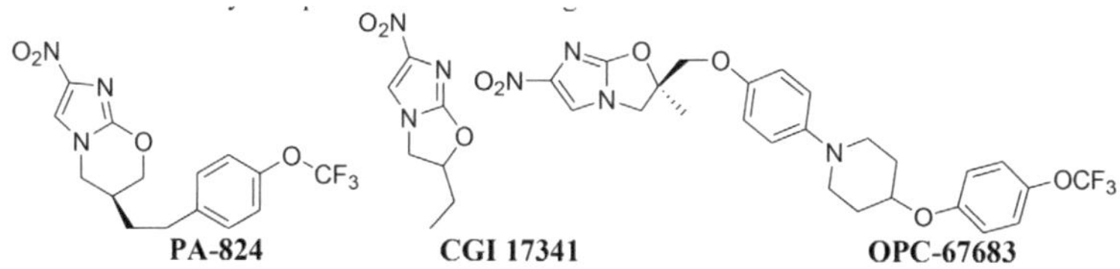

PA-824 **CGI 17341** **OPC-67683**

Imidazole Analogues

Miconazole is a well-known antifungal drug that has been accounted for to have anti-TB activity *in vitro* against *Mtb*H37Ra (MIC 2 μg/ml). It inhibiting replicating bacteria and also has some effect on stationary phase bacilli [88]. Unfortunately, miconazole is not active orally and therefore is little additional interest in progressing TB indication [76].

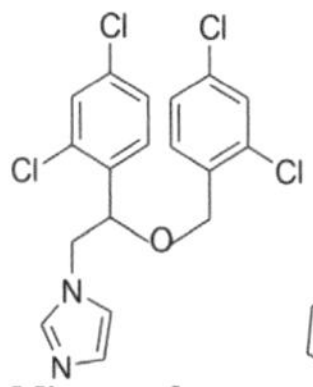

Miconazole

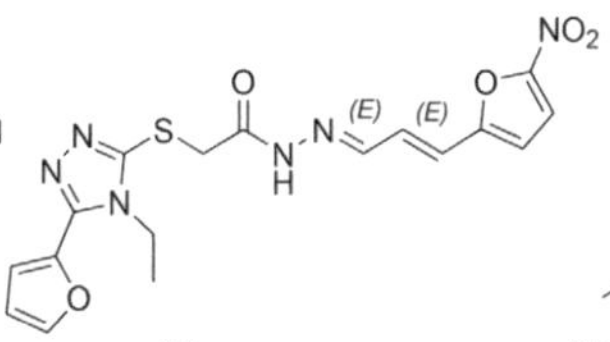

(1)

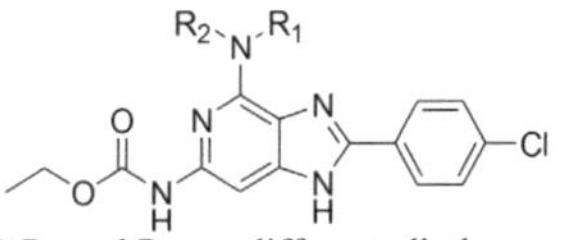

(2) R_1 and R_2 are different alkyl groups

1,2,4-Triazoles

Various 1,2,4-triazoles have been estimated against *Mtb* H37Rv. Compound (**1**) gave 61% inhibition at 6.25 μg/ml. Other triazole analogs were inactive [2–4].

Imidazo(4,5-c)pyridine compounds

A series of imidazo(4,5-c)pyridines, one compound (**2**) for common formula (R_1, R_2 unrevealed)-inhibited *Mtb*H37Rv and other strains with MICs range 0.256–2.56 μg/ml. Imidazo(4,5-c)pyridines were initially prepared as anti-mitotic agents but in the present work, fewer cytotoxic agents were chosen and found to have anti-TB activity [4].

Diarylquinoline compounds

Diarylquinolines (DARQs) is structurally unlike both FQs and other quinolines derivatives. The DARQ R207910 is a new class of anti-TB drugs. It has specificity towards mycobacteria as well as atypical species, important in humans such as MAC, *M. kansai,* and the fast-growing *M. fortium* and *M. abscessus* [77]. The anti-TB specific spectrum differs from that of INH, which has very poor activity against MAC. It will be extremely targeted to the treatment of TB infections, mainly targeting the proton pump of ATP synthase [78].

Diarylquinoline TMC207

Diarylquinoline (DRQ) TMC207 is an exceptionally promising class of anti-TB drugs. About, 20 compounds of the DRQ series have been exhibited a MIC value below 0.5 μg/ml against *Mtb* H37Rv [78]. The most active compound of this class is TMC207 against *Mtb*. The mechanism of action of DRQ TMC207 is different from those of other anti-TB drugs involving a low probability of cross-resistance with accessible anti-TB drugs. The DRQ TMC207 is capable to inhibit bacterial growth against MDR-TB isolates and appears to act by inhibiting the ATP synthase [79], most important to ATP depletion and pH imbalance. Substitutions of RIF, INH, or PZA with DRQ TMC207 hasten activity [80].

Purine Analogs

The 9-Benzylpurines, with a variety of substituents on 2, 6, and/or 8 positions, have high inhibitory activities against *Mtb*. One compound, carrying trans-styryl or aryl substituents at 6 positions and generally chlorine in 2 positions tends to increase the *in vitro* activity and has MIC of 0.78 mg/mL [81]. The anti-TB activity of 6-arylpurines [82] and 9-sulphonylated or sulphenylated-6-mercaptopurines are also known [83].

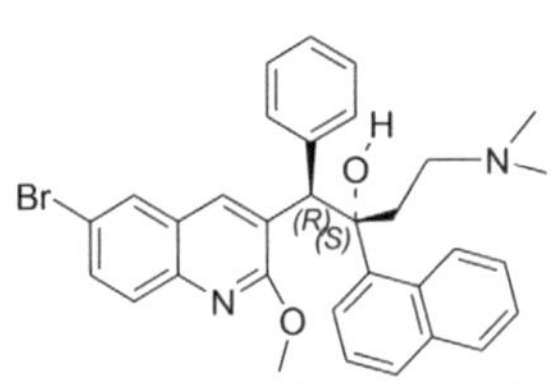

DARQ R207910

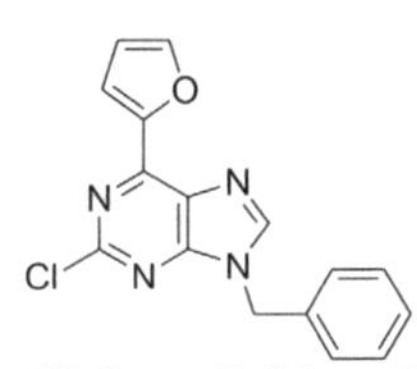

2-Chloro-6(2-furanyl)-9-benzylpurine

Benzylpurines

The 9-benzylpurines, 2-chloro-4(2-furanyl)-9-benzylpurine was potently inhibited *Mtb* H37Rv *in-vitro* with a MIC value of 0.78 μg/ml with low cytotoxicity towards VERO cells (IC_{50} value-8.1 μg/ml) selectivity index (MIC/IC_{50}) of 10.4 [4].

Thiolactomycin Analogues

Naturally occurring (5*R*)-thiolactomycin (TLM) exhibits potent *in vivo* activity against many pathogenic bacteria, and *Mtb* [84]. TLM inhibited bacterial and plant type II fatty acid synthases (FAS-II) but not inhibited mammalian or yeast type I fatty acid synthases (FAS-I) [85]. In *E. coli*, it inhibited both β-ketoacyl-ACP synthase I to III and acetyl coenzyme A (CoA): ACP transacylase activities [86, 87]. The TLM was the first example of naturally occurring thiolactone to displayed antibiotic action. The TLM analogs have improved activity against whole cells of pathogenic *Mtb* strains [88]. The TLM analogs act by the inhibition of the mycolate synthase, an enzyme involves in the biosynthesis of the *Mtb* cell wall.

This has led to the hope that inhibitors of the TLM target enzyme, FAS-II, are potentially important in the treatment of malaria [89], trypanosomiasis, or sleeping sickness [90], and a range of bacterial indications including TB. It also blocks long-chain mycolate synthesis in a dose-dependent mode [91]. The TLM is active *in vitro* against an extensive range of strains of *Mtb*, including INH- resistant, although at somewhat high concentrations. For example, complete inhibition of growth on solid media of the strain *Mtb* Erdman is seen at 25 μg/ml. The TLM itself as an anti-TB agent [92] and racemic mixtures, e.g. compounds (**A**) and (**B**), which are accounted to have superior activity than the parent in inhibiting *Mtb* H37Rv *in-vitro*.

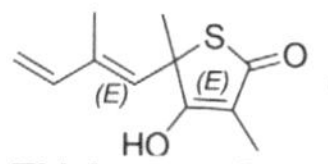

Thiolactamycin

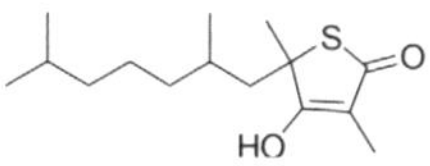

(A)

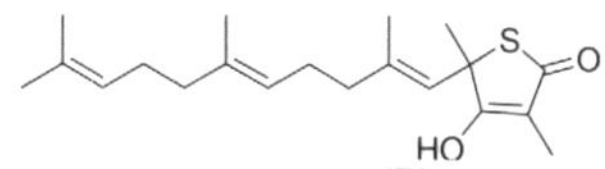

(B)

Mefloquine Analogues

The antimalarial agent mefloquine and its analogs have activity against a range of bacteria including *Mtb* [93]. A series of quinolinemethanol analogs, two compounds, WR-3016, and WR-3017, exhibited potent inhibitory effects in vitro in the *M. avium* complex-1 (MAC) assay with MIC_{50} values of 1 and 2 μg/ml respectively, compared to 16 μg/ml for mefloquine. Other mefloquine analogs, two enantiomers of mefloquine and might be valuable to test some representative 4-aminoquinoline antimalarials such as chloroquine [94]. Another compound desbutylhalofantrine (**3**) is in progress for its antimalarial activities with an advantage over the parent drug halofantrine of lesser cardiotoxicity.

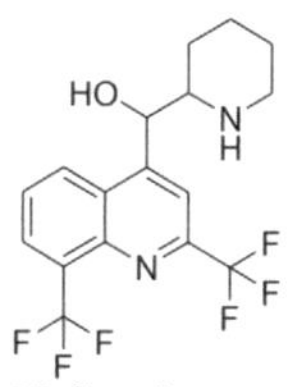

Mefloquine

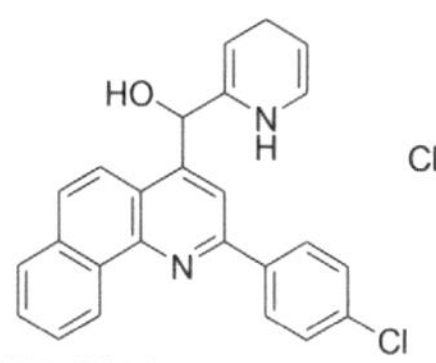

WR-3016

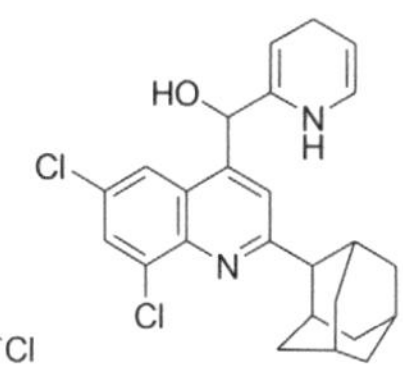

WR-3017

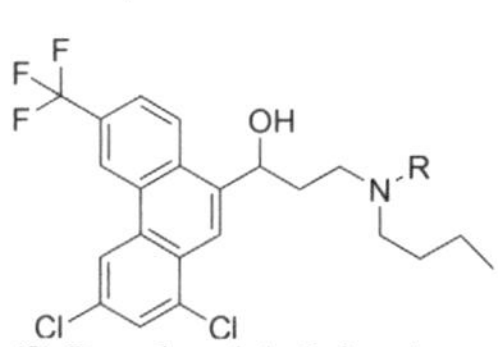

(**3**) R= n-butyl, halofantrine and R= H, desbutylhalofantrine

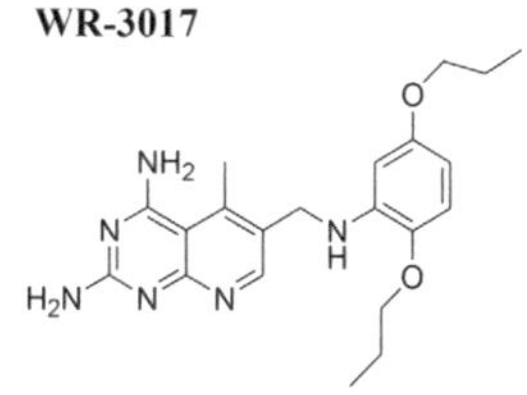

SRI-20094

Deazapteridines

Some 2,4-diamino-5-deazapteridine derivatives, SRI-20094 has potent inhibition of MM6 cells infected with MAC strain NJ3440 with a MIC value 0.13 μg/ml. SRI-20094 inhibits the dihydrofolate reductase (DHFR) of the MAC, with an IC_{50} value 1.0 nM as compared to 4100, 1.4 and 1.0, nM for the trimethoprim, piritrexim, and trimetrexate,. It confirmed limited inhibition for human DHFR having an IC_{50} value 7300 nM. SRI-20094 is a value for the *M. avium* infection and in particular for HIV co-infected persons. Other close analogs were highly active against *Mtb* with MICs of ~0.1 mg/l [95].

1,2,4-Benzothiadiazines

The 1,2,4-benzothiadiazine dioxides have a close relation to sulfonamide and could be considered as cyclic sulfonamides. These compounds exhibited antimicrobial activity [96]. The 1,2,4-benzothiadiazines were explored by incorporating other heterocyclic rings like pyridine and pyrazine moieties (**4** and **5**) and these compounds were exhibited interesting anti-TB activity [97].

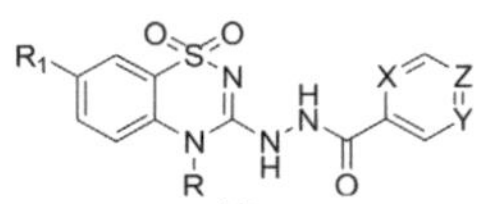

(4)

R= Me, Et, i-Pr, Ph R_1= H, Cl
X=CH, N
Y= CH, N
Z=CH, N, CCl

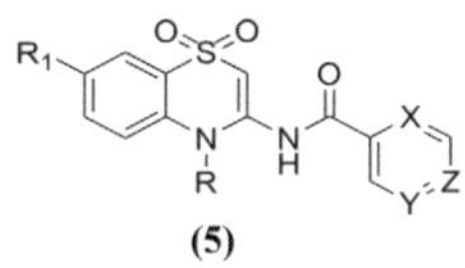

(5)

R= Me,Ph
R_1= H,Cl
X= CH,N
Y=CH,N
Z= CH, N,CCl

Other Molecules

Several other molecules like pyrroles (**6**) [98], quinoxaline-1,4-dioxides (7) [99], and alkylsulfinyl amides (**8**) [100] have been tested for their anti-TB activity. In the analysis of the constant MDR-TB problem, new drugs should concentrate on different targets, including the reduction of TB therapy [101], with negligible toxicity and thus structures based on this lead could provide a novel class of anti-TB drugs.

Benzoxazine derivatives

Some 6-chloro-3-phenyl-4-thioxo-2H-1,3-benzoxazine-2(*3H*)-ones (**9**) and 6-chloro-3-phenyl-2H-1,3-benzoxazine-2,4(*3H*)-dithiones (**10**) have potent anti-TB activity against *Mtb* (MIC values 0.5 mcmol/l), *M. avium* (16 and 16 μmol/l), *M. kansasii* (2 and 2 μmol/l), and *M. kansaii* (1 and 0.5 μmol/l), compared with MIC values of 4, 8, 500 and 500 μmol/l for INH after 14 days [4].

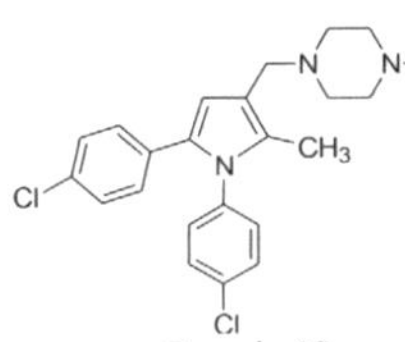

Pyrrole (**6**)

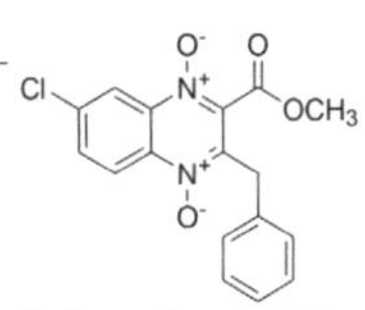

Quinoxaline 1,4-dioxide (**7**)

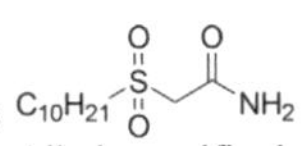

Alkyl sulfinyl amide [FAS20013 (**8**)]

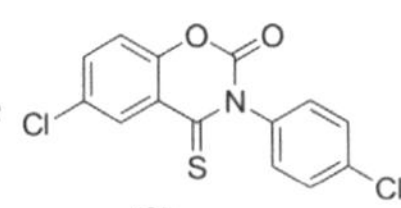

(**9**)

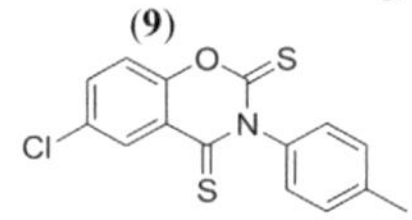

(**10**)

Diterpenoids

Marine products gorgonian coral *Pseudopterogorgia elisabethae* from the West Indian have the anti-TB activity of two active diterpenoid alkaloidal compounds, secopseudopteroxazole and pseudo-pteroxazole [4, 71]. The pseudopteroxazole against *Mtb* H37Rv was claimed to be a powerful inhibitor giving 97% growth inhibition at 12.5 μg/ml even as seco-pseudopteroxazole was rather less active. Some of these derivatives are significantly more active than the marine diterpenoids (11) with the MIC value of *Mtb* H37Rv = 0.46 μg/ml.

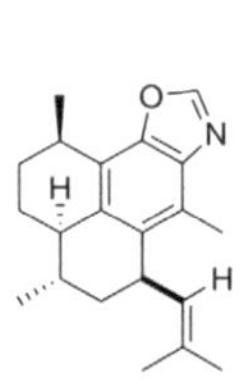

Pseudopteroxazole

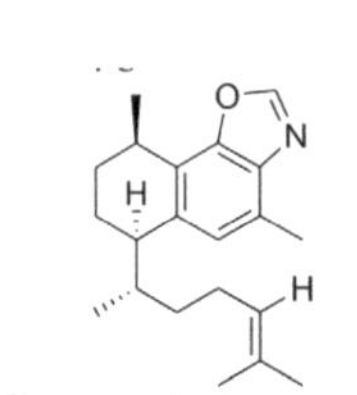

Secopseudopteroxazole

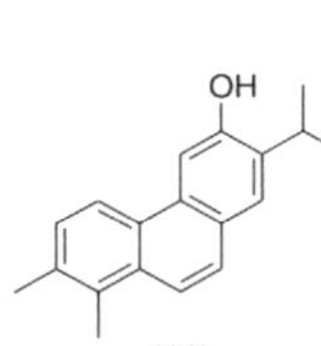

(11)

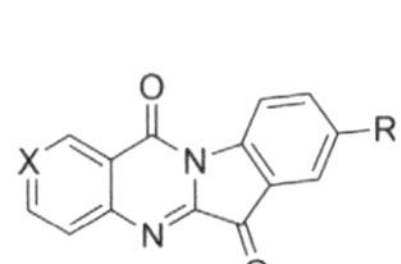

Tryptanthrin
R=H, X=CH & PA505, R=CH(Me)$(CH_2)_5$Me, X=N

Tryptanthrin derivatives

Tryptanthrin is an indoloquinazolinone containing alkaloid and evaluated against different strains of *Mtb* including the drug-sensitive *Mtb* H37Rv strain. The MIC value of tryptanthrin was 1.0 μg/ml compared to the MIC value of INH was 0.03 μg/ml. When evaluated against a section of MDR-TB strains, even as tryptanthrin sustain its effectiveness (MIC = 0.5-1 μg/ml), INH had declined activity with MIC value 4-16 μg/ml. Many derivatives have been tested for their potential in TB treatment like PA-505 having powerful *in vitro* activity towards *Mtb* H37Rv-MIC 0.015 μg/ml and had only modest actions in reducing *Mtb* in the spleen of infected mice when given orally at 50 mg/kg/day for ten days [102].

Clofazimine or Tetramethylpiperidino (TMP) Phenazines Analogues

The tetramethyl piperidine substituted phenazines B4169 and B4128 (TMP phenazines) have possessed significantly activity against *Mtb*, including MDR clinical strains than clofazimines [103]. Recently, new conjugates of phenazine with phthalimido and naphthalimido moieties (**12**) have anti-TB activity [33]. Some phenazine hybrids have shown potential inhibition of *Mtb* ATCC 27294 as well as their clinical isolates (both sensitive and resistant). There is a potential to design new phenazine hybrids for the research and development of new anti-TB agents [14]. The anti-TB effects of tetramethyl piperidinophenazine derivatives are closely related to the clofazimine. The intra- and extracellular effects of these drugs were compared to clofazimine and RIF against *Mtb* H37Rv. The B4169 has effectively inhibited the bacterium with a MIC value of 0.015 μg/ml; the equivalent value for clofazimine was 0.06 μg/ml. These compounds were more active than clofazimine against a series of *Mtb* isolates plus MDR-TB strains. Besides, some derivatives, B4128, exhibited significant intracellular activity (~60% inhibition of growth) at 0.001 μg/ml against *Mtb* infected monocyte-derived macrophages and were better to both clofazimine and RIF drug [14, 15].

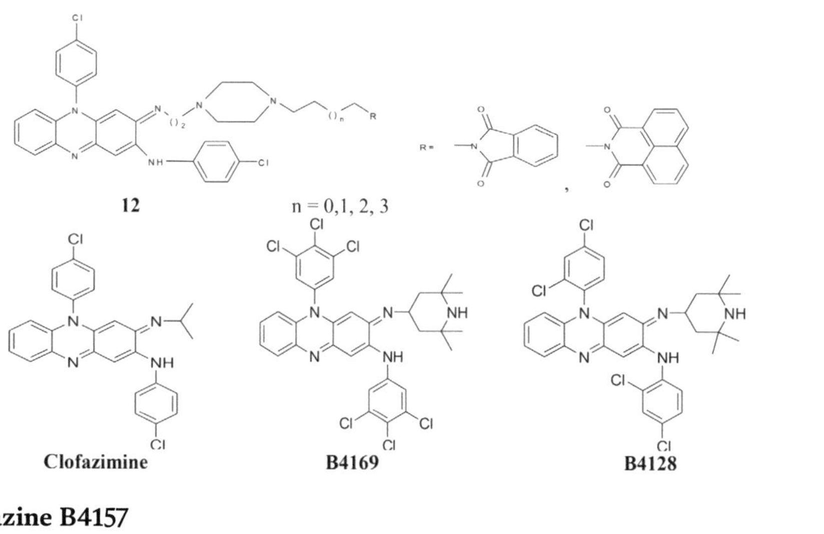

Phenazine B4157

The B4157 is a phenazinamine derivative, closely related to clofazimine, has a potential action for TB. *In vitro*, clofazimine and B4157 were screened against various *Mtb* strains, most of resistance strains, and all were vulnerable to B4157 including which were resistance to clofazimine. The MICs value of B4157 and clofazimine at which 90% of strains inhibited were 0.12 and 1.0 μg/ml. However, C57BL against *Mtb* at 20 mg/kg, clofazimine was slightly better to B4157 [4, 15].

Toluidine Derivatives

Some analogs of toluidines have attractive *in vitro* activity against *Mtb* 103471, the best compound (**14**), having MICs values 4 μg/ml-cf MICs of INH, 0.25 μg/ml, and SM, 0.5 μg/ml. However, these amines will undergo rapid metabolic degradation, possibly toxic metabolites [4, 15].

Saccharides

The arabinose disaccharide SR-9581 is *in vitro* effective against *Mtb*, with a MIC value 4 μg/ml. It reduced the viability of *Mtb* 76.1%, 97.8%, and 99.9% at 8, 16, and 32 μg/ml in 3 days. Another saccharide, an arabinofuranoside oligosaccharide (**14**), substrate for mycobacterial arabinosyltransferases, both compounds can disrupt biosynthesis of *Mtb* cell wall [104, 105].

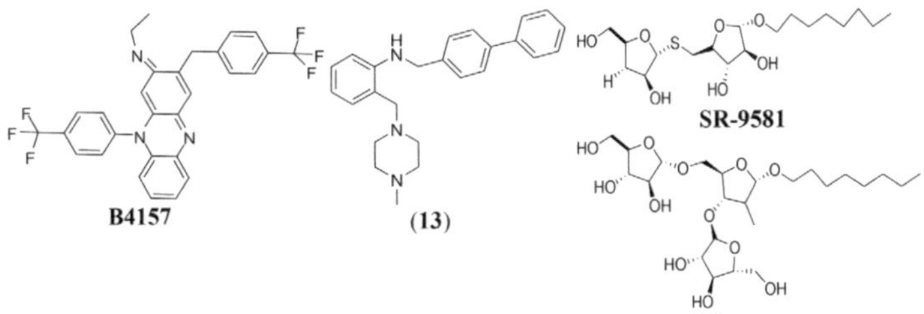

B4157 (13) SR-9581 (14)

Oxazolidinones (Linezolid)

The Oxazolidinones are a class of broad-spectrum antibiotic compounds. They inhibit protein synthesis through binding to the 50S subunit of ribosomes. Oxazolidinones had considerable activities against *Mtb in-vitro* in mice [106]. Oxazolidinones are less promising due to their toxicities and high-cost value [107, 108].

Oxazolidinones PNU 100480 and AZD 2563

They have bacteriostatic activity against various human pathogens together with drug-resistant microorganisms [109, 110]. The oxazolidinones have activity against *Mtb* and linezolid (U-100766) inhibiting MDR isolates *in vitro* at 2 μg/ml [111]. Oxazolidinones having a thiomorpholine group in place of the morpholine group present in linezolid has mainly active against *Mtb* with MICs value 0.125 μg/ml [112]. PNU-100480 was also tested in a murine model against ten strains of *Mtb* in comparison to linezolid and INH. PNU-100480 was found equivalent to INH and more active than linezolid [106].

Linezolid (**U-100766**) X=O PNU & **100480** X=S,

DuP 721 R=MeCO and **DuP 105** R=MeSO

Calanolides

Calanolide A is a naturally pyranocoumarin that has double action against TB and HIV infections. This compound is an inhibitor of the HIV-1 reverse transcriptase enzyme. It also exhibits good in vitro effects towards *Mtb*. In a beginning test of its activity, calanolide A was analogous to the positive control INH and staying effective against RIF and SM resistant TB strains. Calanolide A decreasing the dependency upon acquiring the material from limited natural resources [4, 27] and some compounds, e.g. (**15**), have patented for their anti-TB activities. Calanolide B, which distinct calanolide A, is existing in considerable quantities from renewable natural sources, e.g. from Calophyllum seed oil, [113] has a similar range of activity to calanolide A against *Mtb* and may be an additional cost-effective treatment.

Poloxamer 315 (CRL-1072)

Poloxamer 315 is a methyl oxirane surfactant polymer that shows to disrupt the cell membranes of microorganisms or their intracellular components. The purified polymer is effective against *Mtb* and *M. avium*. *In vitro* effect against *Mtb* showed MIC values 3.1–6.2 μg/ml even as, in a macrophage assay, these go down to 0.92 to 1.25 μg/ml. This compound was effective against strains of *Mtb* resistant to INH, SM, and RIF [114].

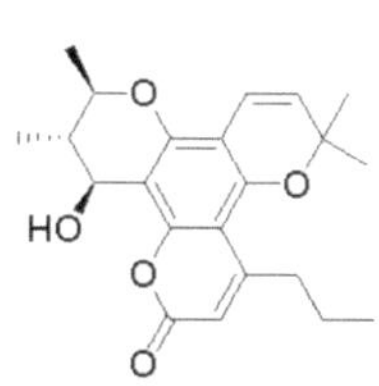

Calanolide A

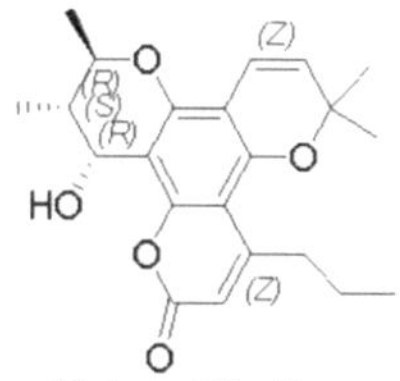

Calanolide B

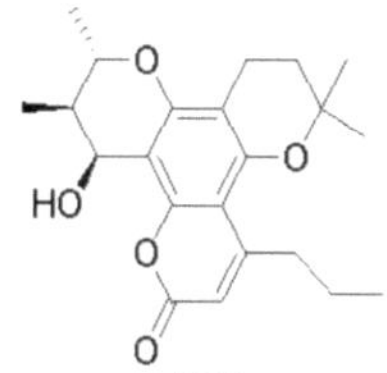

(15)

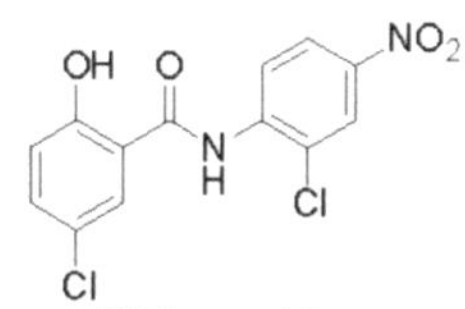

Niclosamide

Niclosamide

Niclosamide

The anthelmintic drug niclosamide was found to have anti-TB activity *in vitro* (MIC 0.5-1 μg/ml) against *Mtb* H37Ra. However, niclosamide is useful for the treatment of human tapeworm infections; it is not absorbed to any significant extent from the intestine [115].

Mikasome

The liposome-encapsulated drug for the anti-TB activity, Mikasome, is useful against *M. avium* infections *in vitro* and *in vivo*. In animals, Mikasome formed 7-fold higher peak plasma levels compared to free drug amikacin (i.v.). The AUC was 150-fold higher with the liposomal substance and a single dose of liposomal amikacin formed therapeutic levels of antibiotics for more than 72 hr. The pilot Phase II studies showed that MiKasome was capable to resolve *Mtb* infections who had failed conventional therapies [4, 14].

Fulleropyrrolidines

A series of fullerene analogs, compound (**2.158**) exhibited anti-TB activity. It inhibited the growth of a human clinical isolate, *Mtb* H6/99, MIC value of 5 μg/ml, and *Mtb* H37Rv MIC value of 50 μg/ml. Some fullerene derivatives have also exhibited *in-vitro* activity against the HIV protease contributing to the tantalizing option of combined actions towards both AIDS and TB [2–4].

Pyrrole LL- 3858

Some pyrroles analogs were effective *in-vitro* against the standard and drug-sensitive *Mtb* strains [116]. Compounds LL-3858 was exhibited a higher bactericidal effect than INH when given as monotherapy to infected mice.

Dipiperidine SQ-609

Dipiperidine SQ-609 is structurally dissimilar to the existing anti-TB drug. It destroyed *Mtb* by interfering with cell wall biosynthesis. The antimicrobial effect has been established *in vivo* in mice models [117–119].

Pleuromutilins

The pleuromutilins is a novel natural antibiotic. They interfere with protein synthesis by binding to the 23S rRNA and consequently inhibiting the formation of a peptide bond [118]. The cross-resistance might happen between pleromutilins and oxazolidinones [119]. Pleuromutilins have been revealed to in-vitro inhibition of the *Mtb* growth. The pleuromutilin compound is active against MDR-TB and permitted shortening of the treatment time.

ATP Synthase inhibitor FAS20013

The FAS20013 belongs to the β-sulphonylcarboxamide analogs. FAS20013 destroys more organisms in a 4-hour exposure than INH or RIF can throughout a

12- to 14-day exposure. This compound is especially effective in killing MDR-TB strains that are resistant to currently used multiple drugs. The greater effect of FAS20013 compared to current anti-TB drugs in terms of its ability to sterilize TB injuries and kill latent TB strains. The FAS20013 has its efficiency in mice with no serious adverse effects and it is up to 100% bioavailable when orally used. The compound is acted by inhibition of ATP synthase [120].

Diamine SQ-109

Diamine SQ-109 was developed as a second-generation drug from the first-line drug ethambutol (EMB). When examined in a low-dose infection model of TB in mice, SQ-109 at 1 mg/kg was as efficient as EMB at 100 mg/kg. However, SQ-109 did not prove improved efficacy at higher doses (10 mg/kg; 25 mg/kg) and was less efficient than INH [121]. The SQ-109 is effictive against MDR-TB, together with those that are EMB-resistant.

Nitrofuranylamides

The *Mtb* is relatively vulnerable to Nitro-containing compounds [122]. Nitrofuranylamide (**16**) was accepted in testing for UDP-Gal mutase inhibition. A prolonged set of nitrofuranylamides was tested for anti-microbial activity. This led to the recognition of several nitrofuranylamides with activity effective against *Mtb* [123].

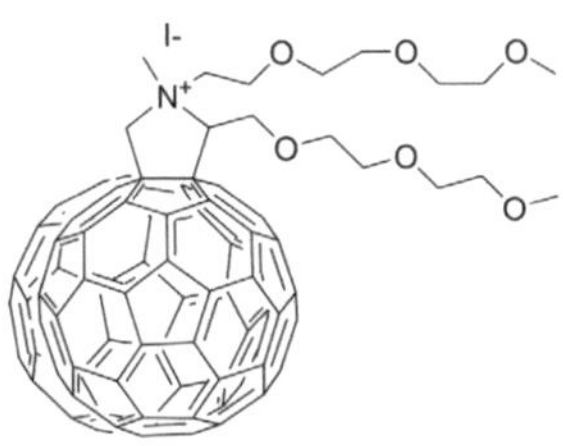

(2.158)

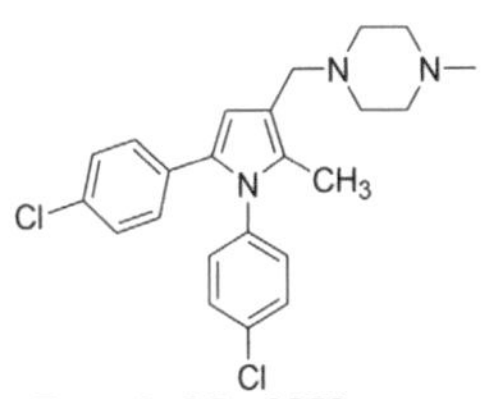

Pyrrole LL- 3858

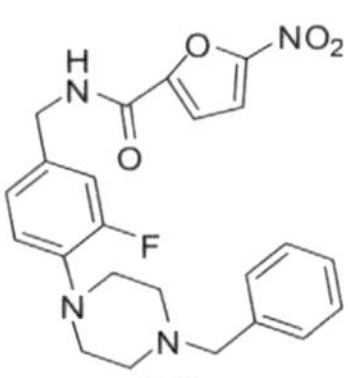

(16)

6. Conclusion

Tuberculosis (TB) is a chronic infectious disease caused by *M. tuberculosis*. Anti-TB drugs developed since the 1940s and their discovery resistance also developed against them. Acquired and primary drug resistances are the common pathways for the development of anti-TB drug resistance. Anti-TB drugs mainly act on protein synthesis, folic acid synthesis, mycolic acid synthesis, DNA synthesis, and ATP synthase. These anti-drugs cause bacteriostatic and/or bactericidal effects on the mycobacterium. The major resistance mechanism is the mutation of the target gene responsible for the action of anti-TB drugs. Anti-TB drug resistance produces a destructive effect on public health. Therefore, the advance study should be conducted in the areas of finding new targets for the development of novel anti-TB drugs.

Author details

Saad Alghamdi[1] and Mohammad Asif[2]*

1 Laboratory Medicine Department, Faculty of Applied Medical Sciences, Umm Al-Qura University, Makkah, Saudi Arabia

2 Department of Pharmaceutical Chemistry, Himalayan Institute of Pharmacy Research, Dehradun, Uttarakhand, India

*Address all correspondence to: aasif321@gmail.com

References

[1] Kebede, B. J. Biomed. Sci., 2019, 8 (3), 1–10.

[2] Asif, M. Orien Pharm & Experi Med, 2012, 12, 15–34.

[3] Asif, M. *Ind. drugs.*, 2012, 49(7), 5–19.

[4] Asif. M. *Mini Rev Med Chem.*, 2012, 12(13), 1404–1418.

[5] Okada, M., Kobayashi, K. Kekkaku, 2007, 82(10), 783–799.

[6] Surendra, S.B., Arya, A., Sudhir, K. S., Vinita, C., Rama, P.T., *Inter. J. Drug Design & Discov.*, 2010, 1(1), 11–18.

[7] Omar, A., Ahmed, M. A. *World Appl.* Sci. J., 2008, 5(1), 94–99.

[8] Elsayed, K.A., Bartyzel, P., Shen, X. Y., Perry, T.L., Zjawiony, J.K., Hamann, M.T. Tetrahed, 2000, 56, 949–953.

[9] World Health Organization (WHO) Global Tuberculosis control epidemiology, strategy, finances. Geneva, Switzerland, 2009.

[10] Health and Social Services, Basic facts about Tuberculosis: TB Control: – Yukon Communicable Disease Control, 2014.

[11] Kampala, T., Shenoi, V.S., Friedland, G. Transmission of Tuberculosis in resource–limited settings. Curr HIV/AIDS Rep., 2013, vol. 10, pp. 3.

[12] Espinal, M. A. Tuberculosis, 2003, 83, 44–51.

[13] Amalio, T., Michael, I. Drugs, 2000, 59, 171–179.

[14] Kamal, A., Azeeza, S., Shaheer, M. M., Shaik, A. A., Rao, M.V. *J. Pharm. & Pharm. Sci.*, 2008, 11(2), 56s–80s

[15] Kamal, A., Reddy, K. S., Ahmed, S. K., Khan, M. N. A., Sinha, R. K., Yadav, J. S., Arora, S. K. Bioorg. & Med. Chem., 2006, 14, 650–658.

[16] Bardou, F., Raynaud, C., Ramos, C., Laneelle, M.A., Laneelle, G. Microbiol., 1998, 144, 2539–2544.

[17] Zhang, Y., Heym, B., Allen, B., Young, D., Cole, S. Nature, 1992, 358, 591–593.

[18] Suarez, J., Ranguelova, K., Jarzecki, A.A. *J.* Biol. Chem., 2009, 284: 7017–7029.

[19] Rawat, R., Whitty, A., Tonge, P.J. Proc. Natl. Acad. Sci., 2003, 100, 13881–13886.

[20] Timmins, G.S., Master, S., Rusnak, F., Deretic, V. Antimicrob. Agent Chemother., 2004, 48, 3006–3009.

[21] Palomino, J.C., Martin, A. Antibiotics, 2014, 3: 317–340.

[22] Wade, M.M., Zhang, Y. Front. Biosci., 2004, 9, 975–994.

[23] Carlos, J.C., Martin, A. Curr. Med. Chem., 2013, 20: 3785–3796.

[24] Piccaro, G., Pietraforte, D., Giannoni, F., Mustazzolu, A., Fattorini, L. Antimicrob. Agents Chemother., 2014, 58, 7527–7533.

[25] Mikusova, K., Huan, H., Yagi, T., Holsters, M., Vereecke, D., et al., J. Bacteriol., 2005, 187, 8020–8025.

[26] Wang, F., Jain, P., Gulten, G., Liu, Z., Feng, Y., et al., Antimicrob. Agents Chemother., 2010, 54, 3776–3782.

[27] Zhang, Y., Yew, W.W. Inter. J. Tuberc. Lung. Dis., 2009, 13, 1320–1330.

[28] Mikusová, K., Slayden, R.A., Besra, G.S., and Brennan, P.J. Antimicrob. Agents Chemother., 1995, 39, 484–489.

[29] Zhang, Y., Wade, M.M., Scorpio, A. Antimicrob. Chemother., 2003, 52, 790–795.

[30] Shi. D., Li, L., Zhao, Y., Jia, Q., Li, H., Coulter, C., et al. *J. Antimicrob, Chemother.,* 2011, 66, 2240–2247.

[31] Shi, W., Chen, J., Feng, J., Cui, P., Zhang. S., et al., *Emerg.* Microbes Infect., 2014, 3, e58.

[32] Zhang, S., Chen, J., Shi, W., Cui, P., Zhang, J., et al. Emerg. Microbes Infect., 2017, 6, e8.

[33] Heifets, L., Desmond, E. Clinical Mycobacteriology Laboratory. In: Cole S, Eisenach K, McMurray D, Jacobs W Jr eds. Tuberculosis and the tubercle bacillus, Washington DC, USA: ASM Press: 2005, 9949–9970.

[34] Chakraborty, S., Gruber, T., Barry, C.E., Boshoff, H.I., Rhee, K.Y. Sci., 2013, 339, 88–91.

[35] Zheng, J., Rubin, J.E., Bifani, P., Mathys, V., Lim, V., et al., J. Biol. Chem., 2013, 288, 23447–23456.

[36] Carette, X., Blondiaux, N., Willery, E., Hoos, S., Lecat–Guillet, N., et al., Nucleic Acids Res., 2011, 40, 3018–3030.

[37] Vanneli, A.T., Dykman, A., Ortiz de Montellano, R.P. *J. Biol. Chem.,* 2002, 277, 12824–12829.

[38] Grant, S.S., Wellington, S., Kawate, T., Desjardins, C.A., Silvis, M.R., et al., Cell Chem. Biol., 2016, 23, 666–677.

[39] Mori, G., Chiarelli, L.R., Riccardi, G., Pasca, M.R. Drug Discov. Today, 2017, 22, 519–525.

[40] Zhang, Y. Annu. Rev. Pharmacol. Toxicol., 2005, 45, 529–564.

[41] Prosser, A.G., Carvalho, S.L.P. Biochem., 2013, 52, 7145–7149.

[42] Pallo–zimmerman, L.M., Byron, J., Graves, T.K. *Compend Contin Educ Vet.,* 2010, 32, 1–9.

[43] Aubry, A., Pan, X.S., Fisher, L.M., Jarlier, V., Cambau, E. Antimicrob. Agent Chemother., 2004, 48, 1281–1288.

[44] Alangaden, G.J., Kreiswirth, B.N., Aouad, A., Khetarpal, M., Igno, F.R., et al., Antimicrob. Agents Chemother., 1998, 42, 1295–1297.

[45] Stanley, R.E., Blaha, G., Grodzicki, R.L., Strickler, M.D., Steitz, T.A. Nat. Struct. Mol. Biol., 2010, 17: 289–293.

[46] Williams, K., Stover, C., Zhu, T., Tasneen, R., Tyagi, S., et al., Antimicrob. Agents Chemother., 2009, 53, 1314–1319.

[47] Bartlett, J.G., Dowell, S.F., Mandell, L.A., File Jr, T.M., Musher, D.M., Fine, M.J. Clin. Infect. Dis., 2000, 31, 347–382.

[48] Grosset, J. H. Tuber. Lung Dis., 1992, 73, 378–383.

[49] Alangaden, G.J., Manavathu, E.K., Vakulenko, S.B., Zvonok, N.M., Lerner, S.A. Antimicrob. Agents Chemother., 1995, 39, 1700–1703.

[50] Ginsburg, A.S., Grosset, J.H., Bishai, W.R. Lancet Infect. Dis., 2003, 3, 432–442.

[51] Ruiz–Serrano, M.J., Alcala, L., Martinez, L., Diaz, M., Marin, M., Gonzalez–Abad, M. J., Bouza, E. *Antimicrob. Agents Chemother.*, 2000, 44, 2567–2568.

[52] Ginsburg, A.S., Hooper, N., Parrish, N., Dooley, K.E., Dorman, S.E., Booth, J., Diener–West, M., Merz, W.G.,

Bishai, W.R., Sterling, T.R. *Clin. Infect. Dis.*, 2003, 37, 1448–1452.

[53] Alangaden, G. J.,Lerner, S.A. Clin. Infect. Dis., 1997, 25, 1213–1221.

[54] Rodriguez, J.C., Ruiz, M., Climent, A., Royo, G. Int J Antimicrob Agents, 2001, 17, 229–231.

[55] Sulochana, S., Rahman, F., Paramasivan, C.N. *J.* Chemother., 2005, 17, 169–173.

[56] Alvirez–Freites, E.J., Carter, J.L., Cynamon, M.H. *Antimicrob. Agents Chemother.,* 2002, 46, 1022–1025.

[57] Paramasivan, C.N., Sulochana, S., Kubendiran, G., Venkatesan, P., Mitchison, D. A. Antimicrob. Agents Chemother., 2005, 49, 627–631.

[58] Cynamon, M.H., Sklaney, M. Antimicrob Agents Chemother, 2003, 47, 2442–2444.

[59] Miyazaki, E., Miyazaki, M., Chen, J. M., Chaisson, R.E., Bishai, W.R. Antimicrob Agents Chemother, 1999, 43, 85–89.

[60] Nuermberger, E.L., Yoshimatsu, T., Tyagi, S., O'Brien, R.J., Vernon, A.N., Chaisson, R.E., Bishai, W.R., Grosset, J. H. Am. J. Respir. Crit. Care Med., 2004, 169, 421–426.

[61] Nuermberger, E.L., Yoshimatsu, T., Tyagi, S., Williams, K., Rosenthal, I., O'Brien, R. J., Vernon, A.A., Chaisson, R.E., Bishai, W. R., Grosset, J. H. Am. J. Respir. Crit. Care Med., 2004, 170, 1131–1134.

[62] Ginsburg, A.S., Sun, R., Calamita, H., Scott, C.P., Bishai, W.R., Grosset, J. H. Antimicrob. Agents Chemother., 2005, 49, 3977–3979.

[63] Burman, W.J., Goldberg, S., Johnson, J.L., Muzanye, G., Engle, M., Mosher, A.W., Choudhri, S., Daley, C. L., Munsiff, S.S., Zhao, Z., et al., Am. J. Respir. Crit. Care Med., 2006, 174, 331–338.

[64] Shoen, C.M., DeStefano, M.S., Cynamon, M.H., Clin. Infect. Dis., 2000, 30(3), S288–S290.

[65] Moghazeh, S.L., Pan, X.S., Arain, T. M., Stover, C.K., Musser, J.M. Antimicrob. Agents Chemother., 1996, 40, 2655–2657.

[66] Andries. K., Verhasselt, P., Guillemont, J., Göhlmann, H.W., Neefs, J.M., et al., Sci., 2005, vol. 307, pp. 223–227.

[67] Koul, A., Dendouga, N., Vergauwen, K., Molenberghs, B., Vranckx, L., et al., Nat. Chem. Biol., 2007, 3, 323–324.

[68] Matsumoto, M., Hashizume, H., Tomishige, T., Kawasaki, M., Tsubouchi, H., et al., PLoS Med., 2006, 3, e466.

[69] Gurumurthy, M., Tathagate, M., Cynthia, D.S., Singh, R., Niyomrattanakit, P., et al., FEBS J., 2012, 279, 113–125.

[70] Stover, C.K., Warrener, P., van Devanter, D.R., Sherman, D.R., Arain, T.M. Nature, 2000, 405, 962–966.

[71] Manjunatha, U.H., Boshoff, H., Dowd, C.S., Zhang, L., Albert, T.J., et al., Proc. Natl. Acad. Sci., 2006, 103, 431–436.

[72] Rao, S.P., Alonso, S., Rand, L., Dick, T., Pethe, K. Proc. Natl. Acad. Sci., 2008, 105, 11945–11950.

[73] Tahlan, K., Wilson, R., Kastrinsky, D.B., Arora, K., Nair, V., et al., Antimicrob. Agents Chemother., 2012, 56, 1797–1809.

[74] Ashtekar, D.R., Costa–Perira, R., Nagrajan, K., Vishvanathan, N., Bhatt,

A.D., Rittel, W. *Antimicrob. Agents Chemother.*, 1993, 37, 183–186.

[75] Sun, Z., Zhang, Y. Tuber. Lung Dis., 1999, 79, 319–320.

[76] Groll, A.H., Walsh, T.J. Curr. Opin. & Infect. Dis., 1997, 10, 449–458.

[77] Shindikar, A.V., Viswanathan, C.L. Bioorg. & Med. Chem. Lett., 2005, 15, 1803–1806.

[78] Andries, K., Verhasselt, P., Guillemont, J., Gohlmann, H.W.H., Neefs, J.M., Winkler, H., Gestel, J.V., Timmerman, P., Zhu, M., Lee, E., Williams, P., de Chaffoy, D., Huitric, E., Hoffner, S., Cambau, E., Truffot–Pernot, C., Lounis, N., Jarlier, V. *Sci.*, 2005, 307, 223–227.

[79] Petrella, S., Cambau, E., Chauffour, A., Andries, K., Jarlier, V., Sougakoff, W. Antimicrob. Agents Chemother., 2006, 50, 2853–2856.

[80] Dolezal, M., Jampilek, J., Osicka, Z., Kunes, J., Buchta, V., Vichova, P. Il Farmaco, 2003, 58, 1105–1111.

[81] Bakkestuen, A.K., Gundersen, L.L., Langli, G., Liu, F., Nolsoe, J.M. *Bioorg, & Med.* Chem. Lett., 2000, 10, 1207–1210.

[82] Gundersen, L.L., Meyer, J.N., Spilsberg, B. *J.* Med. Chem., 2002, 45, 1383–1386.

[83] Scozzafava, A., Mastrolorenzo, A., Supuran, C. T. Bioorg. & Med. Chem. Lett., 2002, 11, 1675–1678.

[84] Miyakawa, S., Suzuki, K., Noto, T., Harada, Y., Okazaki, H. *J.* Antibiotics, 1982, 35, 411–419.

[85] Hayashi, T., Yamamoto, O., Sasaki, H., Kawaguchi. A., Okazaki, H. *Biochem. Biophys. Res. Comm.*, 1983, 115, 1108–1113.

[86] Heath, R.J., White, S.W., Rock, C.O. Lipid Res., 2001, 40, 467–497.

[87] Tsay, J.T., Rock, C.O., Jackowski, S. *J.* Bacteriol., 1992, 174, 508–513.

[88] Douglas, J.D., Senior, S.J., Morehouse, C., Phetsukiri, B., Campbell, I.B., Besra, G.S., Minnikin, D. E. Microbiol, 2002, 148, 3101–3109.

[89] Waller, R.F., Keeling, P.J., Donald, R.G., Striepen, B., Handman, E., Lang–Unnasch, N., Cowman, A. F., Besra, G. S., Roos, D.S., McFadden, G.I. *Proc. Nat. Acad. Sci.,* USA., 1998, 95, 12352–12357.

[90] Morita, Y.S., Paul, K.S., Englund, P. T., 2000. Sci., 288, 140–143.

[91] Slayden, R.A., Lee, R.E., Armour, J. W., Cooper, A.M., Orme, I.M., Brennan, P.J., Besra, G.S. Antimicrob. Agents Chemother., 1996, 40, 2813–2819.

[92] Chambers, M.S., Thomas, E.J. *J. Chem. Soc. Perkin Trans.*, 1997, 1(1), 417–432.

[93] Kunin, C.M., Ellis, W.Y., *Antimicro.* Agents Chemother., 2000, 44(4), 848–852.

[94] De, D., Krogstad, F.M., Byers, L.D., Krogstad, D. *J.* Med. Chem., 1998, 41 (25), 4918–4926.

[95] Li, X.Z., Zhang, L., Nikaido, H. Antimicrob. Agents Chemother., 2004., 48, 2415–2423.

[96] Cocco, M.T., Congiu, C., Onnis, V., Pusceddu, M.C., Schivo, M.L., Logu, A., Eur. J. Med. Chem., 1999, 34, 1071–1076.

[97] Kamal, A., Ahmed, S. K., Reddy, K. S., Khan, M. N. A., Shetty, R. C. R. N. C., Siddhardha, B., Murthy, U. S. N., Khan, I. A., Kumar, M., Sharma, S., Ram, A.B. *Bioorg. & Med. Chem. Lett.*, 2007, 17, 5419–5422.

[98] Deidda, D., Lampis, G., Fioravanti, R., Biava, M., Porretta, G. C., Zanetti, S., Pompei, R. Antimicrob. Agents Chemother., 1998, 42, 3035–3037.

[99] Jaso, A., Zarrana, B., Aldana, I., Monge, A. *J.* Med. Chem., 2005, 48, 2019–2025.

[100] Jones, P.B., Parrish, N.M., Houston, T.A., Stapon, A., Bansal, N. P., Dick, J. D., Townsend, C. A. *J.* Med. Chem., 2000, 43, 3304–3314.

[101] Mitscher, L. A., Baker, W. Med. Res. Rev., 1998, 18, 363–374.

[102] Field, S.K., Cowie, R. L. Chest, 2003, 124, 1482–1486.

[103] Ramneatu, O.M., Lowary, T.L., Poster CARB–52 presented at the 220th National Meeting of the American Chemical Society, Washington D.C., USA, 20–24 August, 2000, Derwent World Drug Alert, abstract WD–2000–011652.

[104] Bertino, J. Jr., Fish, D. Clinical Ther., 2000, 22, 798–817.

[105] Cynamon, M.H., Klemens, S.P., Sharpe, C.A., Chase, S. Antimicrob. Agents Chemother., 1999, 43, 1189–1191.

[106] Brickner, S.J., Hutchinson, D.K., Barbachyn, M.R., Manninen, P. R., Ulanowicz, D.A., Garmon, A., Grega, K. C., Hendges, S.K., Toops, D.S., Ford, C. W., Zurenko, G. E. J Med Chem, 1996, 39, 673–679.

[107] Eustice, D.C., Feldman, P.A., Zajac, I., Slee, A.M., 1998. *Antimicrob.* Agents Chemother., 1998, 32, 1218–1222.

[108] Diekema, D.J., Jones, R. N. Drugs, 2000, 59(1),07–16.

[109] Corti, G., Cinelli, R., Paradisi, F., *Inter.* J. Antimicrob. Agents, 2000, 16 (4), 527–530.

[110] Zurenko, G.E., Yagi, B.H., Schaadt, R. D., Allison, J.W., Kilburn, J.O., Glickman, S. E., Hutchinson, D.K., Barbachyn, M.R., Brickner, S.J., Antimicrob. Agents Chemother., 1996, 40, 839–845.

[111] Barbachyn, M.R., Brickner, S.J. Antimicrob. Agents Chemother., 1996, 40, 839–845.

[112] Spino, C., Dodier, M., Sotheeswaran, S., Anti–HIV coumarins from Calophyllum seed oil. Bioorg. & Med. Chem. Lett., 1998, 8(24), 3475–3478.

[113] Jagannath, C., Reddy, M.V., Kailasam, S., O'Sullivan, J. F., and Gangadharam, P. R., *Am. J. Resp. & Crit. Care Med.*, 1995, vol. 151, pp. 1083–1086.

[114] Cruthers, L., Linenheimer, W.H., Maplesden, D. C., 1979. *Am. J. Veten. Res*, 1979, 40(5), 676–678.

[115] Ragno, R., Marshall, G. R., Di Santo, R., Costi, R., Massa, S., Rompei, R., Artico, M., 2000. *Bioorg.* Med. Chem., 2000, 8, 1423–1432.

[116] Nikonenko, B.V., Samala, R., Einck, L., Nacy, C.A. Antimicrob. Agents Chemother., 2004, 48, 4550–4555.

[117] Kelly, B.P., Furney, S.K., Jessen, M. T., Orme, I.M. Antimicrob. Agents Chemother., 1996, 40, 2809–2812.

[118] Schlunzen, F., Pyetan, E., Fucini, P., Yonath, A., Harms, J. M. Mol. Microbiol., 2004, 54, 1287–1294.

[119] Long, K.S., Poehlsgaard, J., Kehrenberg, C., Schwarz, S., Vester, B. Antimicrob. Agents Chemother., 2006, 50, 2500–2505.

[120] Parrish, N.M., Ko, C.G., Hughes, M.A., Townsend, C.A., Dick, J.D. Effect of noctanesulphonylacetamide (OSA)

on ATP and protein expression in *Mycobacterium bovis* BCG. J. Antimicrob. Chemother., 2004. 54, 722–729.

[121] Protopopova, M., Hanrahan, C., Nikonenko, B., Samala, R., Chen, P., Gearhart, J., Einck, L., Nacy, C. A. *J.* Antimicrob. Chemother., 2005, 56, 968–974.

[122] Murugasu–Oei, B., Dick, T. *J. Antimicrob. Chemother.*, 2000, 46, 917–919.

[123] Tangallapally, R.P., Yendapally, R., Lee, R. E., Hevener, K., Jones, V.C., Lenaerts, A.J., McNeil, M.R., Wang, Y., Franzblau, S., Lee, R.E., J. Med. Chem., 2004, 47, 5276–5283.

Chapter 4

Genealogy of Resistant Tuberculosis in Latin America and the Caribbean until 2020

Onix J. Cantres-Fonseca and Francisco Del Olmo-Arroyo

Abstract

History hallmarks different out brakes events during the last century. Being caught in the in the middle of the catastrophic COVID-19 pandemic, that initiated in 2019 makes possible to forget other causalities. Tuberculosis makes the case. The pathogen has been present more than hundredth years. Relevance rest in worldwide prevalence, pathogen spread, treatment resistance and the need for eradication. Drug treatment resistance is considered as one of the criteria to prioritize a country in the World Health Organization's intention to eradicate tuberculosis infection in the world. For decades in Latin America, including the Caribbean, there have been a persistent high rate of drug resistance with an overall prevalence to one or more drug rounds 13.0%. Approximately 30% of previously treated cases have a multi-drug resistance. In this chapter, we intend to review the epidemiology of resistant tuberculosis, and the causes of resistance associated to the community of people in the Latin American and the Caribbean. We intend to describe the genetic response of *Mycobacterium tuberculosis* from its migratory journey throughout decades from areas of Europa and Asia to Latin America, its genetic transformation secondary to inadequate drug exposure and the characteristics of the infected host, and how a change in the healthcare system and tuberculosis control strategies access are needed to change the surge of multidrug resistance tuberculosis.

Keywords: tuberculosis, treatment, resistance, Latin, America

1. Introduction

More than 100 years have been since the discovery of Mycobacterium tuberculosis (TB) and still continues to be one of the world's leading causes of death by a single infectious organism. Despite extensive knowledge of its pathophysiology and its infectious characteristics, TB continues to cause great morbidity, and continue to be a significant diagnostic challenge. But far from making the diagnosis, the true odyssey occurs when many patients develop an infection with resistance to available therapy.

Mycobacterium tuberculosis is a bacillus shaped, aerobic, slow growing and acid-fast positive staining bacteria. The organism is mostly transmitted through out droplets of particles suspended in the air, and inhalation by the infected host. Deposition of the bacteria in the tissues, especially the lungs, cause the principal manifestations of the disease. Most of the infected people developed a natural immunity by phagocytes which engulf the mycobacteria and form granulomas causing no clinical symptoms. This is known as latent tuberculosis. Differently, around 10% of exposed people develop active infection, characterized by cavitary and fibrotic lung disease [1]. Immunosuppression can cause that tuberculosis bacilli migrate from the initial lung infection to other sites, causing extra pulmonary disease, which is one of the most lethal sequelae of the infection. TB can virtually invade any organ system and mimic other noninfectious conditions, causing to be a significant source of morbidity and mortality [2]. Early identification and appropriate treatment are the clues to obtain a satisfactory outcome.

The diagnosis of TB is done by identification of the mycobacteria in the affected tissue by culture or polymerase chain reaction. Once the organism is identified, combination antimicrobial drug therapy is started, and continued according to drug sensitivity. Length of treatment is determined by the site of infection and the immunologic status of the host, as well as the drug sensitivity of the organism.

Before the end of the first half of the 20th century, antimicrobial medications for Tuberculosis were discovered, following by the use of combination drug therapy as its principal management [3]. Around 1970s, the use of medications as Isoniazid and Rifampin were established as the principal core of treatment that participates in addition to other drugs for treatment of tuberculosis. The combination with other drugs allowed to decrease length of therapy to around 6 months [3]. After more than 50 years, new drugs for tuberculosis have been approved in 2019. The assemblance of previous known drugs and new ones is the pivotal therapy for the infection, including those with resistance to one or more medications (**Figure 1**).

In between the amaze and current focus in the catastrophic pandemic of Covid-19, we tend to forget that a year before in 2018, *M. tuberculosis* infected more than 10 million people, with 1.5 million people dying from the disease [1]. Despite being TB a well-known disease, it is suspected that more than one third of the cases did not receive adequate treatment due to lack of diagnosis and lack of resources [1]. Besides the direct poor outcome in the individual patient, the absence of an appropriate treatment increase the risk of developing drug resistance, facilitates alterations in the genealogy, and favors poor outcomes in morbidity and mortality. Thus, drug resistance is considered a significant menace to populations of high prevalence of disease, and an important set back to eradicate the infection.

Genealogy is described as the study of the history of a specific descendance and how we can follow the different lineages of a family or group. Throughout exposure to drugs, TB has developed the capability of resist antimicrobial therapy, which have evolved in strains that survive beyond those which are sensitive to medications, creating those drug resistant populations of mycobacteria. Current technologies, allow the identification of TB strains with drug resistance using DNA tests which detect genetic mutations to different drugs is less than 48 hours [4]. Also, gene sequencing studies have allowed to follow the lineage of TB in different areas of Latin America. Using specific "gene markers" and mutations detected by

Group	Drugs
First line oral agents	izoniazid (INH) , rifampicin (RIF)
A:Fluoroquinolones	levofloxaxin, gatifloxaxin , moxifloxacin
B:Injectable agents	Streptomycin, amikacin, kanamycin
C: Oral core second line agents	cycloserine, linezolid, ethionamide, protionamide, clofazimine
D: Add on agents	D1: pyrazinamide (PZA), ethambutol (EMB) D2: bedaquiline , delamanid D3: p-aminosalicylic acid, imipenem-cilastin, meropenem, thioacetazone, amoxillin-clavunolanate

Figure 1.
2016 WHO Drug Therapy Groups Tuberculosis.

polymerase chain reactions, the populations of mycobacteria are identified and characterized in different territories [5]. Those genetically identified strains are compared to TB strains in Europa and other continents, and are studied along with migration patterns from those regions, establishing familiar origins.

2. Definition of drug resistance and causes

As mentioned above, the most common therapy for tuberculosis originated from 1970s, which consist in a combination of Isoniazid (INH) and Rifampin (RIF). Adding Ethambutol (EMB) and Pyrazinamide (PZA) can shorten the length of therapy. The first two drugs are considered as the firs line therapy for tuberculosis (**Figure 1**). These are the basic definitions of drug resistant TB:

- **Drug-resistant tuberculosis** is when TB remains unaffected to at least one anti-tuberculosis drug.
- **Mono drug resistance** refers to an infection with resistance to one of the first line agents.
- **Poly-drug resistant TB** occurs when there is resistance to two or more anti-TB drugs but not to both INH and Rifampicin simultaneously.
- **Multidrug-resistant tuberculosis (MDR-TB)** occurs when resistance occurs in more than one antimicrobial drug, or at least isoniazid (INH) and rifampin (RIF).
- **Extensively drug resistance tuberculosis (XDR-TB)**: The extreme case of resistance, in which TB is resistant to at least one drug in each group in the second line therapy groups (see **Figure 1**, groups A to D), besides being resistant to first line therapy.
- For the purpose of this chapter, we will refer and discuss mainly to the multidrug resistant tuberculosis (MDR-TB).

3. Drug resistance

The development of drug resistance of tuberculosis seems to be secondary to a mutation process of a chromosome that can cause that a specific population of mycobacteria develops a "phenotypic resistance" to a certain drug. That mutation can be cause an alteration in the drug transport in the cell membrane of the mycobacteria, or the increase in production of an enzyme that metabolize and cause incapacity of the treatment drug. When those mycobacterias are exposed to either inappropriate antimicrobial therapy, inappropriate length of treatment, poor quality or low dose of medications, or lack of combination therapy, confers to the resistant bacteria a survival advantage that allows a "genetic" transformation which is transmitted over other nonresistant strains [6].

The common use of more than one drug to treat TB resulted from initial studies that showed a progressive increased in mycobacterial populations with resistance in sputum cultures from patients treated only with streptomycin. Combining para-aminosalicylic acid with streptomycin in a clinical trial showed a more than 7 to 8 times decrease in the rate of resistance to streptomycin [7]. With the eventual discovery of the efficacy of other antimicrobial drugs against tuberculosis, multiple other combination therapy studies were done, until current regimes were obtained. When done adequately, many countries have been able to eradicate tuberculosis using direct observe therapy programs monitoring adequate compliance with medications including dosing and regimes.

Each drug against TB has a specific mechanism to cause the bactericidal or bacteriostatic effect on the microorganism. In the case of INH, a specific enzyme is activated by the antibiotic inside the cell to inhibits the production of mycolitic acid, which is integral part of the mycobacterial cell wall [8]. For example, Rifampin inhibits the proliferation of ribonucleic acid, which inhibits genetic material replication and bacterial proliferation.

Mutations that impede Rifampin to inhibit the synthesis of RNA cause drug resistance. In the case of IZH, the drug needs an activation of its initial pro-drug state. A defect in the enzyme that metabolize IZH to its active state will cause resistance to IZH. Rifampin resistance is considered rare and usually occurs concomitantly with resistant of other drugs. This makes the resistance of Rifampin a marker of multiple drug resistance (MDR-TB) [8]. Other drugs as PZA and EMB also function inhibiting the formation of other components of the cell wall, and resistance occurs by mechanism similar to IZH and RIF.

The characteristics of the infected host also can influence in the developing of drug resistance. Immunosuppression allows the mycobacteria to survive the host immunological reaction and proceed to active disease. This can cause population spreading of disease and further dissemination of mycobacterial strains. Immunosuppression include the use of chemotherapy, immunotherapy and anti-inflammatory medications, patients with severe and uncontrolled diabetes mellitus and human Immunodeficiency virus infection (HIV).

Human Immunodeficiency virus infection have been particularly linked to the development of MDR-TB. In patients with HIV there are many factors contributing to the development of drug resistance. Those factors include: the rapid progression of HIV with concomitant tuberculosis infection, poor absorption and interactions between HIV and tuberculosis drugs, and exposure to high risk of resistance populations as other resistant tuberculosis patients, intravenous drug and alcohol users and other patients with poor compliance to medications [9].

Data regarding Diabetes mellitus and development of drug resistance to tuberculosis have been variable. However a recent metanalysis published in 2018, showed significant association between DM and MDR-TB, which was not linked to the degree of economic development of the country. Again, the degree of immunosuppression secondary to uncontrolled diabetes is related to the possibility of developing active disease and infection spreading in the population. The same case occurs with other type of chronic immunosuppression as patients of chemotherapy, immunosuppressive therapy for autoimmune diseases and the chronic use of systemic steroids in patients with respiratory conditions as chronic obstructive pulmonary disease (COPD) and asthma. A publication in 2015 suggested a strong association of MDR-TB with COPD patients with advanced stage disease, severe airway obstruction and long term use of corticosteroids [10]. Also, the multiple use of antibiotics to treat COPD exacerbations, including fluoroquinolones, contribute to the risk of developing drug resistance [10].

4. Epidemiology of drug of drug resistant tuberculosis

It is known that around 4% of newly diagnosed tuberculosis patients are multidrug resistance [11]. From those previously treated, MDR-TB have been reported in more than 20% worldwide [11]. The Global Tuberculosis Report of 2019 recognize more than 400,000 patients with drug resistance TB [1]. In 2019, before Covid-19 became the global threat that we have been seen now, MDR-TB continued to be a public health crisis with 206,030 patients reported with MDR-TB around the world, a 10% increase from numbers reported in 2018 [12]. According to the World Health Organization, the countries with more MDR-TB burden are India, China and Russia.

In America, Center and South American as the areas with more MDR-TB burden. Around 3% of the total tuberculosis cases reported in the world are from American territories [5]. South America was the region with more incidence of tuberculosis with 46.2 per 100,000 of population, mostly from areas of the Caribbean and Central America [5]. The data regarding each region has been variable, but Brazil, Peru, and Mexico have almost half of the cases of Latin America [13]. Costa Rica, Cuba, Jamaica, Puerto Rico, and Trinidad and Tobago reported an incidence close to the threshold for tuberculosis elimination [5].

When talking about MDR-TB in America, previous publications signaled Countries as Peru, Ecuador and Brazil for most of the cases reported [14]. The overall prevalence of MDR-TB according to a metanalysis published in 2020 [15] was 13%, but it reached 28% in those patient populations previously treated cases for tuberculosis. The reported prevalence have been increased when studied from period ending in 2010 to 2018 [15]. Colombia, Mexico, and Dominican Republic, in data published in 2020, have a reported overall prevalence of MDR-TB of around 20% [15], with more of 30% of attributed to a previous failed or inadequate treatment. In the United States, data reports around 1% of MDR-TB of the total of tuberculosis patients identified [16], being considered a low burden country for MDR-TB.

In the Caribbean, as mentioned above Dominican Republic is the country with more prevalence of MDR-TB reported. However, data regarding the number of cases in many of the territories of the Caribbean is scarce. Haiti was also identified as a high burden MDR-TB area, with around 306 cases per 100,000 of population in 2014 [17].

5. Causes of drug resistant in Latin American and the Caribbean territories

Phenotypic resistance of the mycobacteria due to lack of appropriate treatment is considered one of the reasons for which the rate of MDR-TB has not decreased through the years. However, throughout genetic sequencing studies, the genealogy of tuberculosis origin in Latin America has been study. In the case of MDR-TB, considering data that suggest that horizontal gene transfers are nonexistent between *M. tuberculosis* bacteria [5], is genetic transformation probably the cause of widely spread of resistance between the above mentioned populations. In this case, genetic transformation refers to the survival advantage of those strains of mycobacteria with acquired resistance to certain drugs over others more sensitive bacterial population. This occurs when mycobacteria are exposed to inadequate therapy, including low quality and dose of medications, the lack of use of combination therapy, intermittent use of medications or inadequate length of therapy. Data from Peru during the 90's decade, where they stablished Direct Observe Therapy (DOT), showed a significant decrease in the prevalence of MDR-TB [14], consistent with this theorem of phenotypic resistance that culminate in genetic transformation of the mycobacteria in the population.

A recent publication about the genetic epidemiology of tuberculosis in Latin America discuss how the migratory patterns from Europa and Asia have define the genealogy of the mycobacteria in those territories [5]. For example, Peru showed a large number of tuberculosis strains very similar to those found on Beijing, compatible with a well-known historical Chinese migration to that country. Those genetic origins allow the mycobacteria to have different virulence, increasing its capability to survive and being transmitted between hosts. Those mycobacteria have developed the capability to survive the immune response of the host, the external environment in the droplet and particles transmitted, and an increase capability to cause active disease, increasing the probability of population spread. This environmental resistance conferred to those genetically more fitted strains has been associated with an increased risk of developing drug resistance [5].

Again, data is scarce in the specific prevalence of MDR-TB in all regions of the Caribbean. However extremely high incidence of tuberculosis in territories such as Haiti and Dominican Republic have overlap with a high rate of co-infection with Human Immunodeficiency Virus (HIV). The degree of immunosuppression contribute to the developing of active disease, population spreading of infection, and dissemination of selected tuberculosis strains, including those which are drug resistant. High prevalence of other conditions such as COPD and diabetes also contribute to the tuberculosis spread and the development of drug resistant infection. In addition, the limited access and poor quality of health care services, malnutrition, poverty and underprivilege living conditions are associated to the spread of tuberculosis throughout populations [18]. All of those factors are highly prevalent in the Caribbean countries.

6. Conclusion

Limited healthcare and economic resources, inadequate exposure to anti tuberculosis drug regimens and specific characteristics of the infected host, have paved the long way to the genetic inheritance of resistant strains of tuberculosis to develop and propagate in this side of the world. Migratory patterns from Europa and Asia to the Latin American territories have brought tuberculosis strains with genes capable of resists specific drugs, but it has been the inadequate use of medication and the exposures in these lands that have allowed the genetic transformation of bacteria in the populations. As Peru did in decade of 1990, establishing direct observation therapy programs help significatively to decrease the prevalence of MDR-TB strains in the population [14]. Also, prevention of TB spread with rapid diagnosis and screening in high risk populations is a pivotal way to decrease the surge of new cases. The global use of the new genetic rapid tests for detecting drug resistant strains of TB [4] is another way for eradication of MDR-TB. Most important, is to increase population's access to appropriate therapy for those with sensitive and resistant TB infections.

The reach for better healthcare systems, population awareness, programs for better screening and assurance of adequate treatment for tuberculosis patients will be the only infallible tool to decrease the rate of MDR-TB infections and the way to achieve eradication.

Author details

Onix J. Cantres-Fonseca* and Francisco Del Olmo-Arroyo
VA Caribbean Health System, San Juan, Puerto Rico

*Address all correspondence to: onixcantres@gmail.com

References

[1] Maiolini M, et al. The War against Tuberculosis: A Review of Natural Compounds and Their Derivatives. Molecules. 2020 Jul; 25(13): 3011

[2] Cantres-Fonseca O, et al. Extra Pulmonary Tuberculosis: An Overview. Role of Microbes in Human Health and Disease. Intech Open 2019.

[3] Iseman MD. Tuberculosis therapy: past, present and future. European Respiratory Journal. 2002; 20: 87S-94s, DOI: 10.1177/0022219410391190

[4] WHO: Rapid diagnostic test and shorter, cheaper treatment signal new hope for multidrug-resistant tuberculosis patients. May 2016

[5] Woodman M, et al. Tuberculosis Genetic Epidemiology: A Latin American Perspective. Genes (Basel). 2019 Jan; 10(1): 53.

[6] Lallungawi J, et al. Drug resistant Tuberculosis: A review. Comparative Immun Microb and Infect Dis 74. Sept 2020.

[7] Kerantzas CA, et al. Origins of Combination Therapy for Tuberculosis: Lessons for Future Antimicrobial Development and Application. Am Soc for Microbiol; March/April 2017 Volume 8 Issue 2 e01586–e01516

[8] Kurz SG, et al. Drug Resistant Tuberculosis: Challenges and Progress. Infect Dis Clin North Am. 2016 Jun; 30(2): 509-522.

[9] Singh A, et al. Drug-Resistant Tuberculosis and HIV Infection: Current Perspectives. HIV AIDS (Auckl). 2020; 12: 9-31, DOI: 10.2147/HIV.S193059.

[10] Zhao J, et al. Multidrug-Resistant Tuberculosis in Patients with Chronic Obstructive Pulmonary Disease in China. PLoS One. 2015; 10(8): e0135205, DOI: 10.1371/journal.pone.0135205.

[11] Dean AS, et al. Epidemiology of Drug-Resistant Tuberculosis. Strain Variation in the *Mycobacterium tuberculosis* Complex: Its Role in Biology, Epidemiology and Control. pp. 209-22

[12] WORLD HEALTH ORGANIZATION: TUBERCULOSIS

[13] Pan American Health Organisation. Tuberculosis in the Americas 2018. Pan American Health Organisation; Washington, DC, USA: 2018. WHO., DOI: 10.1080/23802359.2018.1483754

[14] Battista Migliori G, et al. Emerging epidemic of drug-resistant tuberculosis in Europe, Russia, China, South America and Asia: current status and global perspectives. Curr Opin Pulm Med 16:171-179, DOI: 10.2471/BLT.14.030514

[15] Tengan FM, et al. Prevalence of multidrug-resistant tuberculosis in Latin America and the Caribbean: a systematic review and meta-analysis. Europen Journal of Tropic & Intern Health, Volume 25, Issue 9; September 2020.

[16] CDC. Fact Sheet: TB Drug Resistance in the U.S. ww.cdc.gov/nchhstp/newsroom/docs/factsheets/tb-drug-resistance-factsheet.pdf. Published October 2015. Accessed June 19, 2017

[17] Millet J, et al. A First Assessment of *Mycobacterium tuberculosis* Genetic Diversity and Drug-Resistance Patterns in Twelve Caribbean Territories. Biomed Research Intern. Volume 2014 |Article ID 718496

[18] Bergonzoli G, et al. Determinants of tuberculosis in countries of Latin America and the Caribbean. Rev Panam Salud Publica 39(2), 2016

Section 3

Latent Tuberculosis Infection

Chapter 5

Latent Tuberculous Infection: Influence on Patient's Quality of Life

Dinara Namazovna Adjablaeva

Abstract

Latent tuberculosis infection is an asymptomatic condition in which patients carry the bacteria, but do not show any sign of illness, however they are at risk of disease activation at any time in the future. Understanding of influence of latent tuberculosis infection on the physical and mental well-being of these patients is important as successful strategies to reduce the tuberculosis burden globally. Our purpose is to explore patients during diagnosis and treatment of latent tuberculosis infection, measure their quality of life. Materials and methods: during 2017–2019 was examined 100 children 4–7 years age. Children were divided in 3 groups. First group (n = 40) - a children with LTI. Group of the comparison (n = 40) has comprised preschool age children with tuberculosis. Group of the checking (n = 20) have constituted the preschool age healthy children. Estimation of children health was conducted by analysis health factors: social, genetic, biological. In addition were studied criteria of health. It was used study anamnestic data, questioning, estimation quality of life, anthropometry, data of objective examination, laboratory data and parameters of functioning, electrocardiography, vegetative nervous system spectrography (VNS-spectrography), manual ergometry. Physical development valued with the help of specialized tables. Leukocyte intoxication index is calculated on formula Shemitova V.F. Variety heart rhythm (VHR) was studied by method VNS-spectrography on vegetotester "VNS-Micro" with computer program "Polispectr" of company "Neyrosoft". Interpretation source vegetative tone and vegetative reactivity was realized according to recommendation N.A. Belokon. Vegetative provision of activity was valued on tolerance to steady-state load by method manual ergometry (MEM) in help of manual dynamometer. Quality of life was defined with the help of questionnaire PedsQL version 4.0 (the Russian version). Results and their discussion: in children with active tuberculosis, specific process has a most negative influence upon quality of life, comparatively temporary negative influence has LTI. Revealed changes in general have brought to reduction of QL indicators both in first and second group. With provision of latency currents of infecting with mycobacteria of tuberculosis, indicators of quality of life should be considered as one of defining, reflecting psychological component adaptation of child, and can be recommended to enter in program of examination and dispensary observation of children with LTI. Conclusions: our study has shown that preschool age children with LTI have rather significant deflections of health condition, revealing by symptoms of intoxication, expressed breaches adaptation and regulation mechanisms. Results of study have logistical confirmed need of improvement of the preventive maintenance and dispensary observation at children with LTI and active

IntechOpen

participation in its base of the interdepartmental approach. All of this allows newly taking a look at problem of the latent tuberculous infection at preschool age children and role general practitioner in preventive maintenance of the development such dangerous diseases as tuberculosis.

Keywords: quality of life, latent tuberculosis infection, treatment, children and adolescents, well-being

1. Introduction

Tuberculosis is a main infectious reason of deaths in the world and one of 10 leading reasons of deaths in the world. From tuberculosis in 2020 have all over the world died 1,85 million people (including four hundred thousand people with HIV), but had suffered 10,4 million people. In 2020 1,3 million children had ill of tuberculosis, and 250 000 children have died from it (including children with HIV-associated by tuberculosis). Tuberculosis is one of the main reasons to deaths of the people with HIV. Serious problem became the tuberculosis with multi drug resistance. However, in global scale number of patients with tuberculosis falls approximately on 2% per annum [1]. Latent tuberculosis infection (LTI) is a state of persistent immune response to stimulation by *Mycobacterium tuberculosis* antigens without evidence of clinically manifested active TB. A direct measurement tool for M. tuberculosis infection in humans is currently unavailable. One-third of the world's population is estimated to have LTI: they do not have active TB disease but may develop it in the near or remote future, a process called "TB reactivation". The lifetime risk of reactivation for a person with documented LTI is estimated to be 5–10%, with the majority developing TB disease within the first five years after initial infection. However, the risk is considerably higher in the presence of predisposing factors.

At present statistics data persuasively show that in some country of the world number of children, for the first time infected with tuberculosis mycobacterium, continues to grow. For instance, in some region to Russia amount such children forms more than 2% whole baby population of the country. Most often primary infecting children of the preschool age, diseases range in this age group on 52% exceeds the general range of diseases amongst baby population [2, 3].

TST (Tuberculin Skin Test) and IGRAs (Interferon-Gamma Release Assays) are the main tests currently available for the diagnosis of LTI. Persons with LTI have negative bacteriological tests: the diagnosis is based on a positive result of either a TST or IGRA test indicating an immune response to Persons with LTI have negative bacteriological tests: the diagnosis is based on a positive result of either a skin (tuberculin skin test, TST) or blood (Interferon-gamma release assay, IGRA) test indicating an immune response to M. tuberculosis. However these tests have limitations as they cannot distinguish between latent infection with viable microorganisms and healed/treated infections; they also poorly predict who will progress to active TB.

Either TST or IGRAs can be used to identify candidates to LTI treatment in high and upper-middle-income countries with estimated TB incidence less than 100 per 100,000. IGRAs should not replace TST in low and other middle-income countries.

Who should be tested and treated for LTI? The risk of progression to active disease is considerably higher in infected individuals who belong to specific high risk populations. Major risk factors for TB activation include: HIV infection, recent contact with an infectious patient, initiation of an anti-tumor necrosis factor (TNF) treatment, receiving dialysis, receiving an organ or hematologic transplantation, silicosis, being in prison, being an immigrant from high TB burden countries, being a homeless person, being an illicit drug user.

Today latent tuberculous infection (LTI) often turn researchers attention as condition, characterized by presence positive tuberculin skin reaction in the background of absence of clinical and roentgenological sign of local (active) tuberculous process. Importance of the problem LTI is in growing of children with such condition. As well as high range of diseases amongst children from this groups speaks that preventive work with infecting by tuberculous mycobacteria children is conducted in insufficient volume [4, 5].

In spite of value of the problem, the role general practitioner in system primary public health care of children with LTI is passive, consists only in discovery of this group children by method tuberculin skin test and issue to phthisiatrician. That has served the cause to persisting study.

2. Purpose of the study

Study health condition of children preschool age with latent tuberculous infection for development of organizing measures on improvement of the preventive maintenance of the tuberculosis and dispensary observation in primary public health care.

3. Problems of the study

1. Study of realized organizing measures on preventive maintenance of tuberculosis and dispensary observation in Uzbekistan and other countries by literature data.

2. Value epidemiological, social and biological risk factors of development infecting by tuberculous mycobacteria of preschool age children with latent tuberculous infection.

3. Reveal the deflections of health and quality of life at children of preschool age with latent tuberculous infection.

4. Develop the organizing measures of preventive maintenance of tuberculosis and dispensary observation of children, infected by tuberculous mycobacteria, in primary public health care.

4. Scientific novelty

For the first time children of preschool age with latent tuberculous infection will are elaborated physician-biological and social factors, which can be reasons low resistivity and infecting with mycobacteria of tuberculosis.

Deflections of health preschool age children will revealed on base clinic-functional complex study with latent tuberculous infection. Will revealed and portioned on groups anamnestic, clinical and laboratory markers of latent tuberculous infection beside of preschool age children.

Will installed influence of latent tuberculous infection on parameters preschool age children's quality of life.

Will scientifically motivated measures of the preventive maintenance of the tuberculosis and dispensary observation for infected children in primary public health care.

5. Materials and methods of study

During 2017–2019 was examined 100 children 4–7 years age on the base of Samarkand state antituberculous sanatorium, kindergarten № 84 of Samarkand.

For including children in conducted study obligatory condition was presence BCG vaccination at birth, attendance by children educational institution.

In study were excluded children, which parents were not agree to participate in study, children with delay psychomotor developments, having chronic diseases with symptoms of intoxication, children, have had sharp disease before 1 month back.

Children were divided in 3 groups. First group (n = 40) - a children with LTI, which were engulfed by dispensary observation and got the chemoprophylaxis in sanatorium. "Sharp turn" of tuberculin skin tests was noted at 3 (7,5%) children, increase of test result on 6 mm and more for one year - at 19 (47,5%), hyperergic result of test – at 12 (30%); annual increase test results with papule size on 12 mm and more - at 3 (7,5%), unchangeable sizes of the test in more than 3 years - at 3 (7,5%) children. Group of the comparison (n = 40) has comprised preschool age children, treated in Samarkand state tuberculosis hospital. Group of the checking (n = 20) have constituted the preschool age children from 1 groups of health with presence correct scar of BCG vaccination. In all group distribution children on sex and age had not a reliable difference.

Estimation of children health was conducted by analysis health factors: social, genetic, biological. In addition were studied criteria of health: physical development, functional condition, level to resistivity, psychomotor development, sharp and chronic diseases in anamnesis, presence of congenital development defects.

It was used study anamnestic data, questioning, estimation quality of life, anthropometry, data of objective examination, laboratory data and parameters of functioning, electrocardiography, vegetative nervous system spectrography (VNS-spectrography), manual ergometry. Information about each child is received information from history of disease and history of child development. Was conducted analysis factor risk of contamination with mycobacteria tuberculosis: physician-biological, social, genetic. Efficiency of vaccination BCG valued on presence scar, size less 4 mm was indicate as faulty vaccination. Physical development valued with the help of specialized tables. Leukocyte intoxication index is calculated on formula Shemitova V.F. Children observed by other specialists.

In help of cluster method all data were generalized. Variety heart rhythm (VHR) was studied by method VNS-spectrography on vegetotester "VNS-Micro" with computer program "Polispectr" of company "Neyrosoft". Registered more than 500 cardiac cycles. Interpretation source vegetative tone and vegetative reactivity was realized according to recommendation N.A. Belokon. Vegetative provision of activity was valued on tolerance to steady-state load by method manual ergometry (MEM) in help of manual dynamometer.

Quality of life was defined with the help of questionnaire PedsQL version 4.0 (the Russian version), for 5–7 age children. Statistical processing was organized on PC Pentium 4. Made descriptive sample method, method one-factorial analysis of variance, Chi-square, U-criterion of Mann-Uitni, factor of Spearmen correlation, criterion of Fisher, reliable were considered differences $p < 0,001$.

6. Results and their discussion

Our study is indicating that at preschool age children with LTI there are deflections of health condition and quality of life. Premorbid background complicated by risk factors of developing tuberculous infection. Amongst specific risk factor, in the main group priority value has a contact with the source of infection (45,0%), presence of the disease at close relatives (42,5%), faulty vaccination BCG - 1-4 mm, absence of the incidence with chemoprophylaxis children with "sharp turn" of tuberculin skin test (17,5%). Reliable difference at frequency specific factor risk with group children, who had treatment in Samarkand state antituberculous sanatorium, was not revealed. Factors, having importance, as at children with LTI, so at children with evident form of the disease, were contact with tuberculous patients and faulty vaccination of BCG.

Important social risk factors at children with LTI were: not working parents (82,5%), asocial lifestyle of family (50,0%), unsatisfactory home conditions (77,5%), alcoholism of parents (72,5%), large families (55,0%). Nearly similar factors detected at children with evident form of tuberculosis. Biomedical risk factors at children with LTI in the first group were polydeficient anemia (35,0%), chronic nonspecific lungs disease (27,5%), preeclampsia and eclampsia during pregnancy in mother's anamnesis (45,0%) - realistically often, than at children with evident tuberculosis. During of cluster analysis are determined as significant factors - early begin artificial feeding (65,0% - on 55,0% often, than in checking group, $p < 0,01$), extensive tooth decay (32,5%, $p < 0,001$), presence to anemia light degree (27,5%, $p < 0,01$) - as at children with LTI, so and at children with evident forms of the tuberculosis. These risk factors promoted lowered resistivity of the children organism, and according to opinion O.B. Nechaeva, could become the reason infecting children with mycobacteria tuberculosis and it persisting in child organism, and speak about insufficiency of mechanism immunological protection. Low resistivity of organism was one of the important sign at group of children with LTI.

At preschool age children with LTI exist the subclinical signs of infecting MBT, basically manifestations of intoxication. Deflections in physical development (PhD) are discovered at 62,5% ($p < 0,001$) of the first group, at each third child from this groups it was connected with low mass of the body (30,0%, $p < 0,01$). Deficit of body mass is probably connected with reinforcement of the processes catabolism, which is called to provide the compensation and adaptation during chronic stressful situation. Intoxication phenomena in the first group children were also expressed as pale skin cover (47,5%, $p < 0,001$), anemia light degree (27,5%), fast heart rhythm (25,0%). These signs were importance as well as in group of the comparison. Intoxication was confirmed by leukocyte intoxication index by V.F. Shemitova. In the main group children this index was positive more than at half of examined person (75,0%).

Condition of peripheral lymph elements in the first group and group of the comparison in principal did not differ, that is indicate of generalities in reactions lymphatic system of children in response to persistence MBT in child organism. However, by comparing these factors in the first group and group of healthy children, noted brightly expressed differences: lymphatic elements dense and elastic consistencies (77,5%), plural (85,0%), size more than 5 mms (85,0%), with perifocal inflammation phenomena (22,5%, $p < 0,01$), unrepresentative localization lymphadenopathy - an cubital (20,0%), parotid (42,5%), occipital (35,0%) in first group was realistically often than in checking group ($p < 0,01$).

Functional systolic noise of heart existed at absolute majority examined children of first (90,0%) and the second group (87,5%). This on 52,5% and 47,5% accordingly often, than at children of checking group (42,5%, $p < 0,001$).

In the first group children with LTI were noted such signs as headaches (27,5%, $p < 0,01$), intolerance of transport (27,5%, $p < 0,01$), pains in chest (17,5%, $p < 0,05$), reduction to concentrations and attention (82,5%, $p < 0,05$), hyperactivity (50,0%, $p < 0,05$), emotional lability (67,5%, $p < 0,05$), petulance (67,5%, $p < 0,05$). During checkup of children with LTI is realistically often discovered marbling of skin cover (100%) and hyperhidrosis of distal extremities (62,5%), than in group healthy children ($p < 0,01$). These signs were significant at group children with LTI, as reasons of the functional breaches to specific organic pathology. Supposedly, these signs can be indicative of sanitary action quality, which conducted in primary section of health care.

Vegetative status at 45,0% children with LTI during cardiography is characterized sympathicotonia (in 20,0% often, than at healthy children, $p < 0,05$), at each fourth child of the main group was noted hypersympathicotonia (25,0%, $p < 0,01$). Considering fact of the final determination of vagus regulation of heart at 5–6 years old and domination of parasympathicotonia at children this age group, sympatic directivity of vegetative regulation, including hypersympathicotonia at children with LTI, follows to consider as overstrain adaptive regulation mechanism on background of chronic stress. It is important to emphasize that sympathicotonia at children of the main group had a relative character and was realized by reason of reduction compensatory of vagal influence that is one of the mechanism of the overstrain adaptational regulation, leaving for frames compensatory and adaptive reactions [5]. Accordingly, neither reliable increase of tension index (TI) (71,0 2,64 u.u.), nor index of Amo at children with LTI (25,6 1,04%), in contrast with checking group (64,5 2,6 u.u. and 30,1 1,2% accordingly), was not received. In this time, was reduced contribution parasympathicitonic influences on initial vegetative status, it was indicated by reliable increasing at children with LTI initial vegetative tonus (IVT) and index of vegetative rhythm (IVR) factors in 12,5% and 15,0% accordingly, in contrast with data of checking group ($p < 0,05$). Such point of view was confirmed identity of mechanism adaptation and regulation of protection at children with chronic infectious disease – tuberculosis. Clinoorthostatic tests has allowed to install that surplus reaction on orthostasis was noted at each fourth child with LTI (25,0%) and at each fifth child with active tuberculosis (20,0%); the normal reaction existed at each fourth child of the main group (25,0%) and only at 5,0% ($p < 0,05$) children of the comparison group. Asympathicotonia of vegetative regulation (AST VR) registered at 50,0% children with LTI - in 1,9 times often, than in checking group ($p < 0,05$). It indicate of essential reduction compensator-adaptive reactivity of vegetative answer at preschool age children of main group, as well as at children with tuberculosis, which AST VR was noted in 2,5 times often (69,2%, $p < 0,05$), than at children of the checking group. Organized cluster analysis has confirmed objectivity and importance of determination VR because revealed AST VR is significant manifestation as at children with LTI, so and at children with active tuberculosis.

Distinctive feature of LTI was safety of normal structure of the wave variability of cardiac rhythm VCR that demonstrated harmony of vegetative provision. Wave characteristic of vegetative spectrum at children of the main group reliable difference, in contrast with data of healthy children, not identified. At children, suffered with tuberculosis, frequency of spectrum VCR was characterized by essential reduction of general power spectrum and all its components due to low compensator parasympathetic influences in structure vegetative balance that was indicative of reduction variability heart rhythm and reserve of powers adaptational mechanism at children with chronic tuberculous infection. Further study has shown that group LTI on spectral feature is not uniform. So at 36,8% children of main group to factors of the spectral features were a similar of healthy children ($M_{healthy} \pm \delta$) - a group "A", but at 63,2% children with LTI wave features of spectrum approached to similar factor of the group children with tuberculosis ($M_{suffered} \pm \delta$) - a subgroup "B". All factors of VNS-spectrum between subgroup "A" and "B" children with LTI were received reliable differences: in subgroup "B" importance of factors total power of the spectrum (TP), waves very low frequency (VLF), low frequency (LF), high frequency (HF), were realistically below on 64,6%, 57,0% and 71,8% accordingly, than at children of subgroup "A". Also, data of children in subgroup "B" were realistically lower, than at children of the checking group VLF on 57,8%, LF - on 58,4%, HF - on 65,6%. Follows to note that factor TP in subgroup "B" was even realistically below (on 19,2%), than at children with active forms of the tuberculosis. At the same time, VNS-spectrum factors between subgroup data "A" and group of healthy children reliable difference was not received. Considering fact, that tuberculosis is an upshot of the early period of primary tuberculous infection, it should be assumed quite justified to performing VNS-spectrography at children with LTI, and if there disorders of spectral wave structure, estimate this as disadvantage trend to disease of tuberculosis and need of taking urgent preventive measures against LTI. Results of spectrogram in examined children shown in **Table 1**.

During of study vegetative provision to activity on data of MEM, determined that tolerance to steady-state load at children with LTI decreased: importance of endurance indicator was 11,6 0,44 (sec), factor of functioning in isometric mode - 143,85,64 (u.u.) - that were on 36,9% and 48,6% less, than at children of checking group. In the main group 44,7% children (on 35,3% more, than at checking group, $r < 0{,}001$) have not were able completely restore of hemodynamic indicators for 3 minutes after performing the test.

Factors of VNS-spectrum	Children group:				p
	Healthy children (n = 20)	LTI (n = 40)		Children with active tuberculosis (n = 40)	
		Having importance $M_{healthy} \pm \delta$ (n = 12)	Not having importance $M_{suffered} \pm \delta$ (n = 28)		
	3	A	B	2	
TP, $мs^2$	8076,0 ± 347,27	7834,0 ± 681,56	2776,0 ± 149,90	3437,0 ± 175,29	$P_{A-B} < 0{,}05$ $P_{A-2} < 0{,}05$ $P_{B-2} < 0{,}05$ $P_{B-3} < 0{,}05$ $P_{2-3} < 0{,}05$
VLF, $мs^2$	1608,5 ± 69,16	1580,9 ± 129,63	679,0 ± 56,71	533,5 ± 27,21	$P_{A-B} < 0{,}05$ $P_{A-2} < 0{,}05$ $P_{B-3} < 0{,}05$ $P_{2-3} < 0{,}05$
LF, $мs^2$	2304 ± 99,10	2221,1 ± 264,32	959,6 ± 82,10	1003,0 ± 51,15	$P_{A-B} < 0{,}05$ $P_{A-2} < 0{,}05$ $P_{B-3} < 0{,}05$ $P_{2-3} < 0{,}05$
HF, $мs^2$	4124,1 ± 177,33	4032,0 ± 479,81	1137,2 ± 184,8	1900,2 ± 96,90	$P_{A-B} < 0{,}05$ $P_{A-2} < 0{,}05$ $P_{B-3} < 0{,}05$ $P_{2-3} < 0{,}05$

Table 1.
Spectrogram data of examined children (M ± m).

Aspects of QL	LTI n = 40 (M ± σ)	Suffered with TB n = 40 (M ± σ)	Healthy n = 20 (M ± σ)
Physical functioning	44,5 ± 2,5	40,2 ± 2,1	88,1 ± 3,4
Emotional functioning	52,1 ± 3,3	53,2 ± 2,4	83,8 ± 3,9
Social functioning	57,2 ± 2,1	42,7 ± 2,1	89,5 ± 2,2
preschool functioning	65,4 ± 2,0	48,0 ± 5,1	77,9 ± 1,3
psychosocial functioning	52,3 ± 2,4	47,9 ± 3,4	83,3 ± 2,8
Total scales	54,8 ± 2,6	46,1 ± 2,4	84,9 ± 2,5

Table 2.
Quality of life indicators in examined groups (in points).

Assessment of life's quality have revealed that at children with LTI quality of life realistically has a low indicators, than at children from checking group, as in opinion of children, as in opinion of tutors.

Indicators of life's quality (QL) in examined groups are presented in **Table 2**.

Indicators of the physical functioning at healthy children were double above than at children with LTI and active tuberculosis - 88,1 ± 3,4 (against 44,5 ± 2,5 and 40,2 ± 2,1 points). Presence of the clinical manifestation of disease is greatly reflected on children's ability to coping with obstacle, run, participation in athletic games. At children with LTI indicators of their physical functioning also were low - 44,5 ± 2,5 that directs that LTI has an influence upon the general condition of organism, that reveals in general weakly expressed malaises at this groups children. This brings them to independent restriction of the daily physical load. At children with active tuberculosis, physical functioning indicators were low. These patients were revealed at peak period of disease that brings sharply expressed change of general condition patients and expressed in practically full refusal of physical loads, daily duties, in accordance with physical activity.

Lowest indicators of emotional functioning noted at children with LTI and active tuberculosis - 52,1 ± 3,3 and 53,2 ± 2,4 points, that indicated of negative influence of tuberculosis to nervous system. Children from this groups more annoyed, moody, whining, feel discomfort from clinical symptoms of disease, at them is often noted presence of alert on cause of contact with persons of opposite sex. Amongst children with active tuberculosis, we have revealed changes in emotional status that is connected with understanding of incurability of diseases, despondency from joining of tuberculosis. High indicators of emotional functioning are registered at healthy children - 83,8 ± 3,9. However, at part of these children is noted presence discomfort from need to visit kindergarten.

During analysis of social functioning highest indicators noted at groups of healthy children – 89,5 ± 2,2 points, and this is indicative of adaptation of children, both to condition of kindergarten, and to acquisition of new friend relationships with other children. In group children with LTI indicators of social functioning were 57,2 ± 2,1 points. This reflects presence of such problems as compelled temporary cessation of the social relations in group in by reason of receiving of preventive treatment in tuberculous sanatorium. Amongst children with active tuberculosis fixed the lowest indicators of social functioning -42,7 ± 2,1 points. As judged by answer respondent of these groups, awe for its future is from realizations contagiousness of diseases, as well as incurability of it.

Preschool functioning practically does not suffer at healthy children - 77,9 ± 1,3 points, only at a part children is revealed by restlessness, absence of attention, as well as inattentive attitude to performing of tasks. At children with active tuberculosis this scale of functioning has a low indicators - 48,0 ± 5,1 points. These patients often skip the occupations in case of its condition, impossibility concentration during occupation. At children with LTI life in school is evaluated of 65,4 ± 2,0 points, the main problems of this group carried temporality - a restriction of visit the kindergarten at period of stay in sanatorium.

Scale of psychosocial functioning is a total scale emotional and social functioning. According results of this scale higher indicators were noted in group of healthy children - 83,3 ± 2,8 points, comparatively low in group children with LTI - 52,3 ± 2,4 points, and realistically low in group of patient with active tuberculosis - 47,9 ± 3,4 points.

Total scale has revealed the regularity - at children with active tuberculosis specific process has a most negative influence upon quality of life, comparatively temporary negative influence has LTI.

Revealed changes in general have brought to reduction of QL indicators both in first and second group. With provision of latency currents of infecting with mycobacteria of tuberculosis, indicators of quality of life should be considered as one of defining, reflecting psychological component adaptation of child, and can be recommended to enter in program of examination and dispensary observation of children with LTI.

Base on the above data introduces following picture of developing deflections mechanism of health condition at children with LTI:

- complex disadvantage physician-social factor leads to chronic stressful situation,
- that provokes adaptive and regulator overstrain and
- leads to immunological insufficiency,
- which clinical revealing low resistivity,
- that in condition not efficient immunization BCG and contact with bacterial isolation,
- is realized as infecting child with tuberculosis mycobacteria, with development LTI,
- it supporting chronic stressful reaction with transition in vicious circle and
- prospect of the failure to adaptation with transition in tuberculosis.

Social disorder connected with breach of main biorhythm regularities of life, high psychoemotional exhaustion on home conflicts background; it brings to reduction of child health and disadaptation of system activity. Certainly, this is an obligate ambience for persisting any infections, in this instance persisting of MBT.

In base of fetter of development LTI at preschool age children, were designed approaches to improvement of preventive maintenance and dispensary observation at participation general practitioner, marked active position of general practitioner in saving of children health, infected with MBT. Also was determinate of favorable current LTI group - "observation" group, overstrains of adaptation - "attention" group and disadaptation - "risk" group; as well as were improved questions of receivership in rendering physician-social help such children between phthisiatrician and family (social institution) with participation primary health care organization. Groups of dispensary observation of children with LTI in polyclinics are presented in **Table 3**.

Functional data	"Observation" group (favorable adaptation)	"Attention" group (overstrain adaptation)	"Risk" group (disadaptation)
Hypersympathicotonia	Absent	Absence/presence	Presence
Spectral features (TP, VLF, LF, HF)	Normal	Reduction	Reduction
Vegetative reactivity	Normal/HST/AST	Normal	HST/AST
System activity	Normal/reduction	Normal/reduction	Normal/reduction/sharp reduction

Table 3.
Groups of dispensary observations at children with latent tuberculous infection in polyclinic.

Consequently, in structure of medical-social help to children with latent tuberculous infection is defined staging of observations. First stage of medical help to children with LTI must be dispensary-policlinic section (district pediatricians, physicians of the educational institutions, general practitioner), which timely revealed children, infected with MBT (by tuberculin skin tests data), and direct to phthisiatrician this children for specialized help. Second stage must be to define the branch a physician-social help, realizing nonspecific rehabilitation of children with LTI, directed on elimination risk factor and consequence chronic physician-social stress. We suppose the expedient observation by general practitioner to children with LTI not less than three years, because long processes of astenization, reduced adaptation require time to value efficiency specific chemoprophylaxis and nonspecific correction. In addition, children with LTI from "attention" and "risk" groups pass the rehabilitation (third stage) on base of general profile sanatorium (more than 1 month). In case of favorable current of adaptation, ("observations" group) rehabilitees can be realized on area, in families or in social institution.

Therefore, our study has shown that preschool age children with LTI have rather significant deflections of health condition, revealing by symptoms of intoxication, expressed breaches adaptation and regulation mechanisms. Results of study have logistical confirmed need of improvement of the preventive maintenance and dispensary observation at children with LTI and active participation in its base of the interdepartmental approach. All of this allows newly taking a look at problem of the latent tuberculous infection at preschool age children and role general practitioner in preventive maintenance of the development such dangerous diseases as tuberculosis.

7. Conclusions

1. Preschool age children with latent tuberculous infection have rather significant deflections in their health condition.

2. Anamnesis of children with latent tuberculous infection greatly burden with specific risk factor of infecting MBT. These are not enough effective BCG vaccination at 72,5% of examined person, tuberculosis at close relative (37,5%), as well as biomedical and social factors, provoking low level of resistance at 55,5% children of main group: early artificial feeding (65,5%), sharp diseases of respiratory tract (42,5%) and asocial family.

3. In spite of latent current of the primary tuberculous infection, at preschool age children there are realistically significant clinical signs: marbling of skin cover

(72,5%), sweating (80,0%), hyperhidrosis (52,5%), deflections of physical development (60,0%), lymphadenopathy (72,5%). At 47,5% children with latent tuberculous infection is defined toxicosis, by first degree of leukocyte intoxication index.

4. Initial vegetative tone at majority preschool age children with latent tuberculous infection identical of healthy children group indicators, with saving sympatic-parasympathetic activity; but at 21,1% children initial vegetative tone similar indicators of children with active tuberculosis and characterized hypersympathic directivity with reduction compensatory influences vagus nerve that reflects significant adaptational tension.

5. Spectral features variety of heart rhythm at 66,0% children with latent tuberculous infection similar data of children with active tuberculosis, which frequency spectrum is characterized by essential reduction to general power of the spectrum (TP) on 57,4% and all its component (LF on 56,5%, HF on 53,9% and VLF on 66,8%). It demonstrate significant overstrain adaptational mechanism of vegetative regulation functional systems of the organism child with latent tuberculous infection.

6. Latent tuberculous infection promotes reduction of adaptability processes of child organism, revealing in deficit of quality of life.

7. Well-timed taking by general practitioner on dispensary register children with latent tuberculous infection, differentiated approach depending on conditions mechanisms of adaptation and regulation, complex and interdepartmental medical examination are contribution to preventive maintenance of development tuberculosis at given contingent of children.

8. Practical recommendation

To pediatrician during estimation of health condition of children with latent tuberculous infection is recommended:

1. Use leukocyte intoxication index by V.F. Shemitova as subclinical markers of intoxication.

2. Use data of heart rhythm variety with determination of indicators: hypersympathicotonia in initial vegetative; the reduction of general power of spectrum and all its components; asympatic vegetative reactivity; reduction to system activity – tolerance to steady-state load and quality of life deficiency as markers of adaptation.

3. Define the groups of dispensary observations of children with latent tuberculous infection: "observations", "attention" and "risk" considering data of adaptation and regulation mechanism: favorable adaptation; the overstrains of adaptation and disadaptation, which

4. Dispensary observation of children is with latent tuberculous infection must be realized in household polyclinic during 3 years in accordance with applicable scheme. Dispensary observation of children with LTI is presented in **Table 4**.

Action	Groups of dispensary observation		
	"Observations"	"Attention"	"Risk"
First year of observation			
Checkup by pediatrician	Once in half year	Once in quarter of year	
Volume of investigation	Assessment of risks, clinical examination, VNS-spectrography, cardiography, manual ergometry (MEM), study of quality of life (QL)		
Specialists checkups	Otophinolaryngology, stomatologist, on evidences - a neurologist, immunologist and others		
	Once in half year or on evidences		
Adaptation and regulation	Enriched feeding, dose physical loads, cleaning of chronic foci of infections		
	Sanitary actions	Energy-metabolic correction, optimizers CNS (by neurologist prescription), immunocorrection (by immunologist prescription)	
Estimation of efficiency	Absence clinic-laboratory and roentgenological manifestations of tuberculosis, positive dynamics of tuberculin skin tests, absence clinic-functional breaches of vegetative regulation and system activity (VNS-spectrography, cardiography, MEM, QL), normalization of physical development, increasing resistivity, improvement of current accompanying pathology		
2 and 3 years of observation			
Checkup by pediatrician	Once in half year		Once in quarter of year
Volume of investigation	Actions of first year observation		
Specialists checkups	Similar as actions of first year observation		
Adaptation and regulation	Similar as actions of first year observation + sanatorium treatment (at least 1 month)		
Estimation of efficiency	Similar as actions of first year observation		

Table 4.
Dispensary observation of children with latent tuberculous infection.

Author details

Dinara Namazovna Adjablaeva
Samarkand State Medical Institute, Samarkand, Uzbekistan

*Address all correspondence to: dinarasammi@mail.ru

References

[1] Nechaeva O.B. Tuberculosis in children in Russia. Tuberculosis and lung diseases. 2020;98(11):12-20. (In Russ.)DOI 10.21292/2075-1230-2020-98-11-12-20

[2] World health organization. WHO Report (2020)

[3] Brown J, Cappoci S. Smith C, Morris S, Abubakar I, Lipman M. Health status and quality of life in tuberculosis. Int J Infect Dis. 2015; 32:68-75. DOI 10.1016/.2014.12.045.

[4] Adzhablaeva D.N. The study of the life quality parameters in children and adolescents with HIV associated tuberculosis. Tuberculosis and lung diseases, 2020;98(9):14-17. (In Russ.) DOI 10.21292/2075-1230-2020-98-9-14-17

[5] Adzhablaeva D.N. Life quality assessment in phthisiopediatrics. Tuberculosis and lung diseases. 2018;96(2):32-35. (In Russ.) DOI 10.21292/2075-1230-2018-96-2-32-35

Section 4

Molecular Characterization MTB

Chapter 6

Molecular Characterization of *Mycobacterium* spp. Isolated from Cattle and Wildlife in Poland

Anna Didkowska, Monika Krajewska-Wędzina, Blanka Orłowska, Monika Kozińska, Ewa Augustynowicz-Kopeć and Krzysztof Anusz

Abstract

Although Poland is officially tuberculosis (TB) free, meaning that less than 0.1% of her cattle herd is TB-positive, the problem of bovine TB in Poland may be re-emerging: its presence has recently been confirmed in domestic and companion animals, wildlife such as the European bison, and even humans. The aim of this chapter was to review all reports of bovine TB in Poland described to date, with particular emphasis on molecular studies, and determine further research directions. These studies include a range of molecular methods for diagnosis, including genotyping, spoligotyping and MIRU- VNTR; such methods successfully identifies a tuberculosis-positive European bison as the source of wild boar infection in the Bieszczady Mountains based on its spoligotype. This chapter argues that identified trains should be better archived, as such records would allow detailed epidemiological investigations and shed greater light on the activity of *Mycobacterium* spp. The current epidemiological situation in Poland highlights the need for further studies to determine epidemiological links and confirm possible routes of transmission based on whole genome sequencing; this need is accentuated by the zoonotic potential of such infections and the endangered species at risk.

Keywords: epidemiological investigation, European bison, molecular methods, *Mycobacterium caprae*, *Mycobacterium bovis*, Poland

1. Introduction

Bovine tuberculosis is a highly-contagious bacterial disease whose etiological agents are the acid-fast bovine mycobacteria species *Mycobacterium bovis* and *Mycobacterium caprae*. These two species can also cause tuberculosis in humans, although infection with *Mycobacterium tuberculosis* is more common. Although Poland has been an officially tuberculosis free (OTF) country since 2009, cases are still noted in cattle. In addition, *M. bovis* infection has been observed in emergency cases in alpacas and *M. caprae* has been found in endangered European bison (*Bison bonasus*). Tuberculosis infection has been observed in various other wildlife. The aim of this chapter is to present the epizootic situation of bovine tuberculosis in

IntechOpen

Poland, including its molecular diagnostics, and to determine which molecular diagnostic methods would be useful in the future.

2. Tuberculosis in Poland as a zoonosis

Tuberculosis (TB) remains a leading cause of death worldwide. Its treatment requires supervision, efficient and reliable diagnostics, contact tracing and effective therapy. In 2019, 10 million people with tuberculosis were registered worldwide. The incidence of tuberculosis in Poland is slightly higher than the European average, being 13.9/100.000 in 2019 [1, 2].

Although most cases of human TB are caused by the bacterial species *M. tuberculosis*, this only represents part of a complex that includes various zoonotic forms. According to the latest nomenclature, some of the most prominent members of this complex known to cause disease in humans or/and animals are *M. tuberculosis* (human), *Mycobacterium africanum* (human), *Mycobacterium canetti* (human), *M. bovis* (cattle and other animals), *M. caprae* (goats, cattle and other animals), *Mycobacterium pinnipedii* (seal), *Mycobacterium microti* (voles and other small rodents) and *M. bovis* BCG (vaccine strain) [3, 4].

While transmission can take place directly, through the aerogenic route, bovine tuberculosis (bTB) is most commonly transmitted to humans though an indirect route, possibly through unpasteurized milk or dairy products and raw meat. Those at the highest risk of indirect exposure are people exposed to the source of infection at work, such as farmers and veterinarians, and those working with meat, such as slaughterhouse workers and hunters in contact with contaminated animals [5].

According to estimates by the World Health Organization (WHO), in 2016, 147,000 new cases and 12,500 deaths were associated with zoonotic tuberculosis worldwide. However, such figures are often underestimated due to financial constraints and the consequent lack of adequate routine control in countries where bovine tuberculosis is endemic. Zoonotic tuberculosis tends to be of low prevalence where its presence is correctly monitored in animals and appropriate safe food production procedures are followed [6, 7]. While over two thirds of human TB cases, i.e. those resulting from *M. tuberculosis* infection, primarily affect the lungs [8], zoonotic TB often affects extrapulmonary sites, including lymph nodes and other organs [9]. Since bovine mycobacteria causes clinical, radiological and pathological symptoms that are similar to *M. tuberculosis*, these strains can be distinguished only by bacterial culture, by biochemical and morphological analysis, and by genotyping.

M. bovis used to be differentiated from other complex members based on its resistance to pyrazinamide (PZA); however, following the discovery of PZA-susceptible strains of *M. bovis*, the species was split into two subspecies: the PZA-resistant *M. bovis* subsp. *bovis*, and the PZA-sensitive *M. bovis* subsp. *caprae* [10, 11]. PZA is one of the four essential drugs used in the current standard first-line anti-TB treatment regimen. However, as most healthcare providers initiate treatment without performing any drug susceptibility testing, patients with zoonotic TB caused by *M. bovis* may demonstrate poorer treatment outcomes and may develop further resistance to other anti-TB drugs; for example, additional resistance to rifampicin and isoniazid has been detected in some *M. bovis* isolates [12].

Like other bacterial species, the resistance of *Mycobacterium tuberculosis* complex members to antimycobacterial drugs arises from the selection of naturally-resistant mutants that are constantly present in every bacterial population. Wild strains of mycobacteria belonging to the *M. tuberculosis* complex that have never been exposed to drugs are naturally sensitive to tuberculostats, with one exception; the PZA-resistant *M. bovis*.

In addition to mutations, mycobacteria can develop phenotypic resistance through a change in cell wall permeability, which can impair penetration of the drug into the cell, or by employing efflux pumps, which allow the active removal of the drug from the cell. Metabolic pathways that bypass "drug-sensitive" sites in the cell may also be altered. Regardless of its mechanisms, mycobacterial drug resistance always occurs as the result of a selection process, i.e. a change in the ratio of drug-sensitive to drug-resistant cells [13, 14].

With the growth of research of the *Mycobacterium* genome and its drug resistance pathways, the main mechanisms and genes by which mutations cause drug resistance have been recognized. Currently, commercial tests based on PCR reactions are used to detect the most common mutations determining the resistance of mycobacteria to antibiotics that are crucial in treatment. One such example is the line probe assay (LiPA); these employ targeted amplification of specific regions of the MTB genome using biotinylated primers followed by reverse hybridization of the amplicons to oligo probes immobilized on nitrocellulose strips. Hybridization is then detected by a colorimetric reaction. Currently, the most widely used tests are those developed by Hain Lifescience (Nehren, Germany): Genotype MTBDRplus and Genotype MTBDRsl, detecting resistance to the most important antituberculosis drugs of the I- and II- lines [13].

From the epidemiological and therapeutic point of view, the identification of MDR (Multi Drug Resistant), XDR (eXtremely Drug Resistant) and TDR (Totally Drug Resistant) MTBC strains is of key importance [15, 16].

Drug-resistant tuberculosis is more difficult to treat than drug-resistant tuberculosis. Patients do not recover from the standard six-month treatment regimen, but undergo long-term therapy requiring the use of less effective, more toxic and more expensive drugs (**Table 1**) [17].

In Poland and most other developed countries, the threat of bTB in humans decreased significantly in the middle of the 20th Century following the introduction of tuberculosis management strategies [18]. Thanks to the combined implementation of appropriate eradication and surveillance programs, in 2009, Poland was awarded the status of a tuberculosis-free country.

However, in 2020, the first Polish case of bTB in humans was recorded in a retrospective study by Kozińska and Augustynowicz-Kopeć [19], which described the case of a 46-year-old male detected in 2012 with bacteriologically-confirmed pulmonary infection with *M. caprae*. Changes typical for pulmonary TB were

Drug	Type of resistance
Isoniazid + rifampicin	MDR
Isoniazid + rifampicin + streptomycin	
Isoniazid + rifampicin + ethambutol	
Isoniazid + rifampicin + streptomycin + ethambutol	
MDR + fluoroquinolone + one of the injectable drugs (amikacin or kanamycin or capreomycin)	XDR
INH + RMP + SM + EMB + fluoroquinolone + aminoglycoside + polypeptide + thioamide + cycloserine + para-aminosalicylic acid	TDR

Table 1.
Definitions of drug resistance on MTBC.

identified on chest X-ray, and a tuberculin test result of 18 mm was obtained. In addition, direct staining of sputum revealed the presence of acid-fast mycobacteria (AFB ++), and *Mycobacterium* colonies were identified after four weeks of culture on Löwenstein-Jensen (LJ) medium. Initial identification in the hospital laboratory confirmed that the isolated strain belonged to the *M. tuberculosis* complex. Phenotypic and molecular methods revealed drug susceptibility, and the strain was thus classified to the species *M. caprae*. Further genotyping identified the unique spoligotype 200003757377600; although this strain was not registered in the international spoligotype databases SpolDB4 and SITVIT WEB, it was found to match SB1690 in Mbovis. Org, this being a Spanish isolate from 2009 (**Table 2**) [20]. The source of infection remained unknown: the patient's history revealed that he had not had recent contact with any person with tuberculosis, nor had he been close to farm animals which had not been tested for tuberculosis. Until now, this has been the only documented case of zoonotic TB in Poland.

As tuberculosis is an infectious disease with a complex epidemiology and pathogenesis, it is essential to employ molecular typing (genotyping) methods when testing for *M. tuberculosis*: such tools are fundamental for guiding effective epidemiological research, defining the dynamics of transmission, and enabling global surveillance of the disease. In addition, genotyping provides an insight into the biodiversity and evolution of the pathogen.

Various genotyping methods are used in human and bovine TB research, such as IS*6110*-RFLP (*Insertion Sequence 6110-Restriction Fragment Length Polymorphism*), spoligotyping, MIRU-VNTR (*Mycobacterial Interspersed Repetitive Units-Variable Number Tandem Repeats*), and WGS (*Whole Genome Sequencing*) [21].

The spoligotyping method takes advantage of a polymorphism within the chromosomal region DR (*Direct Repeat*) found in mycobacteria belonging to the *M. tuberculosis* complex. This region, first described by Hermans in the *M. bovis* BCG P3 strain, is formed by a variable number of direct repeat (DR) sequences, 36 bp long, with short (35–41 bp) unique spacer sequences between them [22]. The spacer sequences are detected by synthetic oligonucleotide probes complementary to the 43 known sequenced spacer sequences identified in *M. tuberculosis* H37Rv and *M. bovis* BCG strains.

Being a PCR-based method, spoligotyping requires very little DNA and thus, can be used to detect and identify *M. tuberculosis* complex bacteria directly in clinical specimens, bypassing the culture step.

Another advantage of spoligotyping is the ease with which typing results can be recorded, i.e. in binary and octagonal formats, cataloged, and compared in central

Results of microbiological and molecular testing	
Clinical material	Sputum
Bacterioscopy	++
Culture	Growth after four weeks on LJ medium
Phenotype	Sensitive to SM, INH, RMP, EMB, PZA
Strain identification	*Mycobacterium caprae*
Spoligotyping	Hybridization pattern □■□□□□□□□□□□□□□□■■■■■■□■■■■□■■■■■■■■■■□□□□□

Table 2.
Characteristics of Mycobacterium caprae *– The first human isolate in Poland.*

databases [23] (SpolDB4, SITVIT WEB, Mbovis. Org databases). It is therefore commonly employed as a screening method in molecular epidemiological investigations. It can be used to identify species within the *M. tuberculosis* complex, provide information regarding the lineage of various strains, determine their placement in major genetic families and indicate the directions of the global spread of molecular families of mycobacteria [24].

However, the detection of tuberculosis transmission foci in closed populations of humans and animals, as well as their interspecific transmission, requires the use of methods with a higher genome differentiation potential. Therefore, spoligotyping studies are commonly complemented by the use of MIRU-VNTR analysis and WGS [25].

The largest group of VNTR sequences in the *Mycobacterium* genome are the 46–100-nucleotide MIRU fragments. Of these 15 to 24 known *loci* with the highest variability were selected for the genetic typing of mycobacteria.

In the MIRU-VNTR method, individual sequences are amplified, and the size of the resulting products depends on the number of repeats of the core unit. For each locus, the number of repeats of the MIRU or VNTR motif is calculated, which allows the results to be cataloged using a 15- or 24-digit MIRU-VNTR code. The MIRU-VNTR method is characterized by high sensitivity and repeatability. It allows the analyzed strains to be differentiated to a large extent, is relatively easy and is distinguished by a short analysis time [26].

Although spoligotyping, MIRU-VNTR and RFLP have a very high diagnostic value, they are not suitable for accurately determining the dynamics of TB transmission. The spread of tuberculosis may also occur through short contacts, or in a high-risk population where epidemiological links between patients are difficult to establish. In addition, as they screen less than 1% of the genome, standard genotyping techniques therefore have limited discriminatory power and cannot optimally detect potential transmission chains.

These limitations can be circumvented by the use of whole genome sequencing (WGS). WGS provides comprehensive genetic data as well as information on drug resistance, virulence factors, and genome evolution. However, such sequencing analysis requires high expenditure, the possession of specialized equipment and complex bioinformatic analysis of the results [27].

An accurate confirmation of the molecular relationship of the studied strains, supplemented with epidemiological data, can form the basis for identifying the transmission of infection between closely-related patients, such as family members, as well as among homeless people and immigrant populations, between wild animals and livestock, and between humans and animals. Unfortunately, not all diagnostic laboratories have the appropriate equipment to perform specialist testing based on the analysis of the mycobacterial genome. As a result, current data on the transmission of tuberculosis as zoonosis may well be underestimated.

Preventing the development of zoonotic TB in humans requires reducing the risk of exposure and transmission at the human-animal interface. However, while the principal routes of transmission are known, more information is needed about their underlying sociocultural and economic bases, and how to promote safer alternatives.

3. Epizootic situation of bovine tuberculosis in cattle and other animal species in Poland, and the molecular characteristics of isolated strains

Bovine tuberculosis is an infectious disease that mainly affects cattle. In 2020, seven outbreaks in cattle were recorded in Poland; in the rest of Europe, only France (n = 105) and Germany (n = 10) reported higher numbers of outbreaks, while seven outbreaks were noted in Italy and Belgium [28]. Bovine bacilli can cause tuberculosis in other farm species (**Figure 1**). They show high virulence in natural conditions in goats, pigs, sheep and cats [29]; however, the disease is less common in horses and dogs [30, 31] Cattle are not very susceptible to human bacilli, but infections with *M. tuberculosis* are known in this species: one case of bovine tuberculosis due to *M. tuberculosis* has been reported in Poland so far [32]. Wild animals living in the close vicinity of farms can also be a mycobacterial reservoir. The largest reservoir of bovine bacilli in Great Britain is the badger population [33]. However, in Spain, wild boar populations represent the largest reservoir of tuberculosis [34]. The transmission of tuberculosis bacilli occurs in shared pastures, less often as a result of fighting or biting.

In Poland, the largest reservoir of bovine bacilli is believed to be sick cattle. The spread of infection between herds is usually due to the movement of asymptomatic vector animals. Introducing infected animals into a tuberculosis-free herd may cause infection of other animals and disease development in immunocompromised animals. However, following the eradication program carried out in Poland in 1959–1975, its prevalence has significantly fallen, especially in the eastern part of the country. Further progress in the control of the disease in cattle herds has been made possible by the application of strict rules and their consistent enforcement. As in other European countries, Poland operates a special bovine tuberculosis control program, described in detail in the Regulation of the Minister of Agriculture and Rural Development and in the amended Instruction of the General Veterinary Inspector. These documents require the testing of 1/5 of the total cattle population in each county based on bovine and avian purified protein derivative (PPD) tuberculin using both single and comparative tuberculin tests. All positively reactive animals are eliminated, and all samples from these animals are tested in the National Reference Laboratory of Bovine Tuberculosis, located in the

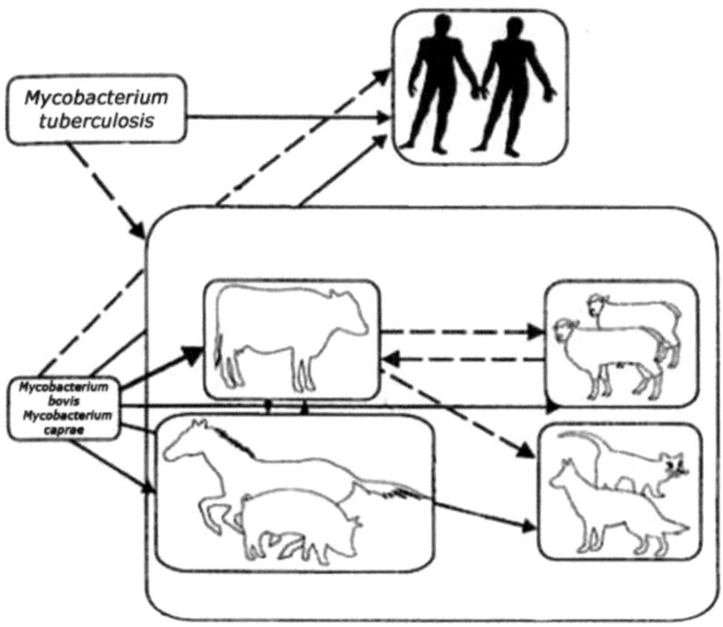

Figure 1.
Diagram illustrating transmission of potential tuberculosis cases caused by mycobacteria from the MTBC complex.

Department of Microbiology of the National Veterinary Research Institute (NVRI) in Puławy, Poland.

All tissue samples are collected *post mortem*, prepared and cultured on Stonebrinck and Petragnani media, as stipulated by the State Veterinary Inspectorate's Instructions for Laboratory Diagnosis of Bovine Tuberculosis [35]. The culture is also supplemented by a biological analysis performed using two guinea pigs, and the strain extraction procedure is complemented by GenoType MTBC (HAIN Lifescience, Germany) typing kits.

The numbers of outbreaks and sick animals in cattle breeding were found to fall during the course of the program, and Poland was recognized as free from bovine tuberculosis in 2009. However, a total of 372 outbreaks were recorded in cattle herds during the following 10-year period, i.e. in 2009–2019. Almost 1/3 of these outbreaks were found in the Masovian Voivodeship, the central region of Poland, especially in its northern part [36]. A significant number of outbreaks were also recorded in the Greater Poland (Wielkopolskie) (n = 68), Lodzkie (n = 28) and Lesser Poland (Malopolskie) voivodeships (n = 24). The smallest number of disease outbreaks concerned the Lubusz (Lubuskie) (n = 2) and Opole voivodeships (n = 1). Molecular studies to date indicate that 70% of cattle suffering from bovine tuberculosis were infected with *M. bovis* and 30% with *M. caprae*. In 2010, the first case of tuberculosis in a calf caused by *M. tuberculosis* was confirmed [33]. It should be noted, however, that the status of Poland as being officially free of bovine tuberculosis was never threatened; on the contrary, compared to other European countries, Poland has very favorable data on disease control, particularly considering that almost six million cattle are farmed there.

Among the *M. bovis* strains isolated from cattle in Poland, the most common individual spoligotype was SB0856, being present in 44% of the tested strains. In addition, SB0127 and SB0119 were also frequently observed [37]. Among *M. caprae* strains isolated from cattle, the most common spoligotypes were SB0418, SB2390 and SB2393 [38].

However, the incidence of bTB is not limited to cattle. In the period of 2009–2010, cases of bTB were recorded in three zoos in Poland. Of the 12 strains isolated from 12 captive animals, *viz.* six antelopes, three giraffes, two tapirs and one alpaca, those from ten animals were identified as *M. bovis* and two were identified as *M. caprae* [37, 39, 40]. Transmission was only confirmed in the antelope herd and between tapirs. Unfortunately, it has been suggested that zoos may withhold epidemiological data, making it very difficult to conduct epidemiological investigations and trace the source of infection among rare and valuable animals threatened with extinction. During a tuberculosis outbreak in the Slaski Ogród Zooologiczny (*Silesia Zoo*), the decision was made to treat a giraffe with active tuberculosis, with unfortunately negative results [41]; this raises the question of whether the research team should have undertaken the treatment of an animal with active tuberculosis, particularly when considering the potential consequences for public health.

In 2014, a bovine tuberculosis outbreak was also identified among American bison (*Bison bison*) farmed in Poland [42]. In total, three cases of *M. caprae* strains were isolated, all of which were characterized by the spoligo pattern SB1912. Most cases of TB among free-living animals and exotic animals kept in zoos in Poland are caused by *M. caprae* strains, with spoligotypes SB1912, SB2391 and SB2392 predominating [38].

In contrast, sporadic cases of transmission to other species of livestock and domestic animals have been reported. For example, one case was found in pigs (*Sus scrofa* f. *domestica*) kept in the vicinity of a herd in which advanced disease was diagnosed [43]. In 2018, bovine tuberculosis was reported among alpacas of British origin in Poland [44]. In both cases, the strain was identified as *M. bovis* spoligotype

SB0666, according to the international spoligotype database (www.Mbovis.org); this type was first isolated in Great Britain in 2003. bTB has also been confirmed in both free-living wild animals and those in breeding centers [45].

Animal strains of MTBC have been analyzed for drug resistance to five basic anti-tuberculosis drugs: streptomycin (SM), isoniazid (INH), rifampicin (RMP) and ethambutol (EMB), known as SIRE, and PZA. fortunately, the findings indicate that Polish strains of bTB obtained from animals do not show environmental resistance [38, 40, 46].

A gap exists in Polish veterinary legislation regarding bovine tuberculosis: so far, it makes no explicit mention of *M. caprae* causing tuberculosis in animals. In the Act of 11 March, 2004 on the Protection of Animal Health and Control of Infectious Diseases of Animals, Annex 2, bovine tuberculosis is listed as a notifiable disease without a disease-causing pathogen. While the disease is mentioned in the Regulation of the Minister of Agriculture and Rural Development of 23 November 2004 on eradication of bovine tuberculosis, it does not indicate an etiological agent. Despite the Amendment of the Instruction of the Chief Veterinary Officer No. GIWpr-02010/2016 of 8 February 2016, the only pathogenic species listed as causing bovine tuberculosis is *M. bovis*.

Poland was declared OTF in 2009 [47], and the fact that the country has remained this way for the subsequent 10 years indicates that the procedures used to control the disease are effective. Only minor incidents have been reported, and they usually occur as a result of incidental errors in anti-epizootic management and the carelessness of animal owners. More importantly, such errors do not seem to have a decisive impact on the overall bovine tuberculosis situation. Poland currently has a consistent policy of eradicating *M. bovis*/*M. caprae* infections in cattle herds, and the country still meets the formal requirements for a TB-free status.

4. Bovine tuberculosis in European bison in Poland and the use of molecular methods

Even though bTB-positive cattle are considered to constitute the primary reservoir of the bovine mycobacterium in Poland, tuberculosis has also been found in wildlife such as badgers (*Meles meles*), wild boar (*Sus scrofa*), wolves (*Canis lupus*) and European bison (*Bison bonasus*) [45, 48, 49].

In recent years, of all species diagnosed with bTB in Poland, the European bison is the most common [50]. A total of 45 cases of tuberculosis were confirmed in European bison in the Bieszczady Mountains during the years 1996–2013 [51]. An autopsy identified generalized tuberculosis in a three-year-old female from a free-living herd in the Brzegi Dolne Forest District. Around the same time, in the years 1997–2001, 13 out of 18 culled European bison from the same *Brzegi Dolne* herd were microbiologically confirmed to have tuberculosis and the decision was made to liquidate the entire herd [49]; however, not all animals were culled, and several bison from the herd have still not been found [51].

Other scattered cases have been found in the region. Tuberculosis was confirmed in two European bison in the Bieszczady Mountains in 2005–2008 [52]. In addition, a positive result in the *Górny San* herd from Bieszczady in 2009 resulted in the entire herd of 24 European bison being culled. Tuberculosis-like lesions were found in all individuals, and tuberculosis was microbiologically confirmed in 23 [51]. It is possible that the source of infection for the European bison from the *Brzegi Dolne* herd was locally grazed cattle, while the source of infection in the *Górny San* herd may have been individuals that separated from the *Brzegi Dolne* herd. Unfortunately, as no strains from the *Brzegi Dolne* herd were archived, it was not possible to compare the mycobacteria strains between the two herds; this underlines the importance of using molecular methods when studying epidemiology among wildlife. Interestingly, a strain isolated from the *Górny San* was found to have the same spoligotype as one isolated from wild boar from the same area, i.e. the Bieszczady Mountains [49].

Cases of bTB have been recorded in captive European bison in Poland: in Warsaw Zoo, Wolisko and the Smardzewice Bison Breeding Centre. Spoligotyping and MIRU-VNTR analysis of the European bison from Smardzewice identified the presence of as *M. caprae*–spoligotype *M. bovis* _4_ CA 1600 (octagonal pattern: 200003770003600) (SpolDB4 database) [53]. The source of infection remains unknown due to a lack of archived *Mycobacteria* strains, but there are suspicions that it may have been acquired from an individual from *Silesia Zoo*.

A number of studies have been undertaken recently to address the problems associated with the *ante mortem* diagnosis of tuberculosis in wildlife [54–56]. Such studies have also been conducted in European bison [57]. Although a range of serology methods have been tried [58], the material for direct detection is collected from tracheobronchial lavage, and from swabs and biopsy from retropharyngeal lymph nodes. A more recent approach is to combine microbiological testing with molecular tests, allowing accurate results to be obtained in a much shorter time. In one case, MTBC genetic material was confirmed in laryngeal swab and tracheobronchial lavage using the BD ProbeTec *Mycobacterium tuberculosis* Complex (DTB) Direct Detection Reagent Pack (Becton Dikinson, US) which allows direct detection of mycobacterial genetic material in a clinical specimen [57]. The test acts by amplifying and identifying the target DNA simultaneously. However, the method is characterized by *inter alia* intermittent mycobacterial shedding, which can lead to false positive results.

With the current situation of bTB in European bison in Poland in mind, it would clearly be advisable to include molecular methods in routine diagnostics, thus facilitating more accurate epidemiological investigations and more effective disease control.

5. Tuberculosis in wildlife in Poland, other than European bison, including molecular diagnostic methods

Currently, no wildlife tuberculosis monitoring program exists in Poland, except when visible lesions suggestive of TB are found in the animal. Despite this, it seems that tuberculosis cases are rarely found in wildlife in Poland and are limited to the area of the Bieszczady Mountains in Southeast Poland: a region bordered by Slovakia and Ukraine, with the highest peak being Tarnica (1346 m a.s.l.). This area is characterized by high forest coverage, low human population and low livestock abundance [59], unpublished data of the County Veterinary Inspectorate, Ustrzyki Dolne, Sanok]. Between 1996 and 2020, most TB cases in this area were found in European bison and in wild boar [49, 51, 53, 60–63], and no cases have been reported in domestic animals or livestock since 2005. Outside this region, only two single cases of TB have been described in wildlife in Poland: the first in a roe deer (*Capreolus capreolus*) near Gdańsk and the second in a European bison in Borecka Forest [64, 65].

In the Bieszczady Mountains, the first TB case in wildlife was described in 1996 in a European bison from the Brzegi Dolne Forest District [52]. Between 1997 and 2013, TB was recorded in a total of 40 European bison in the region, resulting in the culling of two bison herds (*Bison bonasus caucasicus*) (see section 4) [37, 39, 49, 52, 60, 66–68]. Since then, no new TB cases have been detected within this species in the Bieszczady region [63]. Even so, over the past 20 years, TB has been found in other species of wild animals in the Bieszczady Mountains, mostly in wild boars.

The first case of TB in a wild boar was reported in 2012 in a four-year-old female from Nasiczne in the Bieszczady, which was found dead due to *Metastrongylus* spp. invasion. Postmortem examination showed small caseous, yellowish tubercules in submandibular lymph nodes, from which *M.caprae* was isolated (at that time *M. bovis* ssp. *caprae*). The strain was found to have the same spoligo pattern as those strains isolated in 2011 from European bison from the Bieszczady area [45], this being 200003777377400, or SB2391 as assigned by www.Mbovis.org [69].

Since then, a number of cases of TB have been found in the Bieszczady wild boar population each year. Between 2012 and 2017, *M. caprae* was isolated from the lymph nodes of 21 out of 55 investigated wild boar [63]. These strains were subjected to molecular analysis based on spoligotyping according to Kamerbeek et al. [70], and MIRU-VNTR typing, as given in the public protocol [71]. A total of 15 loci were investigated: MIRU 4, MIRU 10, MIRU 16, MIRU 26, MIRU 31, MIRU 40, VNTR 424, VNTR 577, VNTR 2165, VNTR 2401, VNTR 3690, VNTR 4156, VNTR 2163b, VNTR 1955 and VNTR 4052. All 21 isolated strains shared an identical spoligotype 200003777377400 – SB2391. From this group, 19 strains shared a single MIRU-VNTR pattern (464652364413423), while the other two had patterns that differed with regard to a single locus (464552364413423 and 463652364413423) [63].

To describe the occurrence of TB in wildlife other than European bison and wild boar, both within the Bieszczady Mountain region and elsewhere, lymph node samples were collected for analysis from red foxes, wolves, badgers, red deer, roe deer and brown bear between 2011 and 2017. *M. caprae* was isolated from the lymph nodes of one roe deer and three wolves. Those animals had no visible TB-like lesions [48, 63].

All molecular research of *M. caprae* strains isolated from wildlife in the Bieszczady Mountains has been performed based on hsp65 sequence analysis, the GenoType®MTBC (Hain Lifescience, Germany) test, spoligotyping and MIRU-VNTR analysis. Further studies to determine the epidemiological link and the possible route of transmission of the source of infection are needed based on whole genome sequencing.

6. Conclusions

In conclusion, bovine tuberculosis remains a real threat in Poland, as indicated by the increasing number of cases observed in wildlife and the recent report of the first confirmed case of *M. caprae* infection in human. *M. caprae* is the main etiological agent of bovine tuberculosis in wildlife, and *M. bovis* in cattle.

We recommend that in Poland, bovine tuberculosis should not only be monitored in cattle but also in wildlife. This is especially true in the European bison population, which seems to be highly sensitive to infection. This is highly important for protecting public health, maintaining the OTF status of Poland and of course, protecting the European bison themselves. In which case, particular attention should be paid to the free-living animal population in the Bieszczady Mountains.

There is also a particular need to monitor alpacas, as TB-positive animals pose a particular risk to children and disabled people due to increased contact during animal therapy.

We recommend the more intensive use of molecular tests in monitoring and the proper archiving of the identified DNA. Such molecular methods play an essential role in epidemiological investigations, as these can accurately identify the source of infection and effectively control the disease. Their findings also allow steps to be taken to reduce the spread of infection. Further studies would be of particular value in this regard, particularly those based on whole genome sequencing of archived strains of *M. bovis* and *M. caprae* from different species in Poland.

Conflict of interest

The authors declare no conflict of interest.

Acronyms and abbreviations

AFB	acid-fast mycobacteria
bTB	bovine tuberculosis
DRs	direct repeat spacers
EMB	ethambutol
INH	isoniazid
LiPA	Line probe assays
LJ	Löwenstein-Jensen
MDR	multidrug resistant
MIRU-VNTR	mycobacterial interspersed repetitive units-variable number tandem repeats
MTBC	*Mycobacterium tuberculosis* complex
OTF	officially tuberculosis free
PPD	purified protein derivative
PZA	pyrazinamide
RFLP	restriction fragment length polymorphism
RMP	rifampicin
RR	rifampicin resistant
SIRE	streptomycin, isoniazid, rifampicin, ethambutol
SM	streptomycin
TDR	totally drug resistant
TB	tuberculosis
WGS	whole genome sequencing
WHO	World Health Organization
XDR	extremely drug resistant

Author details

Anna Didkowska[1*], Monika Krajewska-Wędzina[2], Blanka Orłowska[1], Monika Kozińska[3], Ewa Augustynowicz-Kopeć[3] and Krzysztof Anusz[1]

1 Department of Food Hygiene and Public Health Protection, Institute of Veterinary Medicine, Warsaw University of Life Sciences (SGGW), Warsaw, Poland

2 Department of Microbiology, National Veterinary Research Institute, Puławy, Poland

3 Department of Microbiology, National Tuberculosis Reference Laboratory, National Tuberculosis and Lung Diseases Research Institute, Warsaw, Poland

*Address all correspondence to: anna_didkowska@sggw.edu.pl

References

[1] Global Tuberculosis Report. WHO, 2020

[2] Tuberculosis and Lung Diseases in Poland in 2019. Editor: Korzeniewska-Koseła M. Tuberculosis and Lung Diseases Research Institute. Warsaw, 2020

[3] O'Reilly LM, Daborn CJ. The epidemiology of *Mycobacterium bovis* infections in animals and man: a review. Tuberculosis and Lung Diseases. 1995;76:1-46. DOI: 10.1016/0962-8479(95)90591-x

[4] Riojas MA, McGough KJ, Rider-Riojas CJ, Rastogi N, Hernando Hazbón M. Phylogenomic analysis of the species of the Mycobacterium tuberculosis complex demonstrates that *Mycobacterium africanum*, *Mycobacterium bovis*, *Mycobacterium caprae*, *Mycobacterium microti* and *Mycobacterium pinnipedii* are later heterotypic synonyms of Mycobacterium tuberculosis. Int J Syst Evol Microbiol. 2018;68:324-332. DOI 10.1099/ijsem.0.002507

[5] Langer AJ, LoBue PA. (2014). Public health significance of zoonotic tuberculosis caused by the *Mycobacterium tuberculosis* complex. Zoonotic Tuberculosis: *Mycobacterium bovis* and Other Pathogenic Mycobacteria. 3rd ed. Wiley; 2014. 21-33. DOI:10.1002/9781118474310

[6] Roadmap for Zoontic Tuberculosis. WHO, OIE, FAO and The Union, 2017

[7] Global Tuberculosis Report. WHO, 2017

[8] Sandgren A, Hollo V, van der Werf MJ. Extrapulmonary tuberculosis in the European Union and European Economic Area, 2002 to 2011. Eur. Surveill. 2013;18:20431

[9] Durr S, Müller B, S Alonso, Hattendorf J, Laisse CJ, van Helden PD, Zinsstag J. Differences in primary sites of infection between zoonotic and human tuberculosis: results from a worldwide systematic review. PLoS Negl Trop Dis. 2013, 7(8):e2399. DOI: 10.1371/journal.pntd.0002399

[10] Aranaz A, Liebana E, Gomez-Mampaso E, Galan JC, Cousins D, Ortega A, Blazquez J, Baquero F, Mateos A, Suarez G, Dominguez L. *Mycobacterium tuberculosis subsp. caprae* subsp. nov.: a taxonomic study of a new member of the *Mycobacterium tuberculosis* complex isolated from goats in Spain. International Journal of Systematic Bacteriology. 1999;49:1263-1273. DOI: 10.1099/00207713-49-3-1263

[11] Niemann S, Richter E, Rusch-Gerdes S. Differentiation among members of the *Mycobacterium tuberculosis* complex by molecular and biochemical features: evidence for two pyrazinamide-susceptible subtypes of *M. bovis*. Journal of Clinical Microbiology. 2000;38:152-157

[12] Rivero A, Márquez M, Santos J, Pinedo A, Sánchez MA, Esteve A, et al. High rate of tuberculosis reinfection during a nosocomial outbreak of multidrug-resistant tuberculosis caused by *Mycobacterium bovis* strain B. Clin Infect Dis. 2001;32:159-161. DOI: 10.1086/317547

[13] Miotto P, Zhang Y, Cirillo DM, Yam WC. Drug resistance mechanisms and drug susceptibility testing for tuberculosis. Respirology 2018;23(12):1098-1113. DOI: 10.1111/resp.13393

[14] Smith T, Wolff KA, Nguyen L. Molecular Biology of Drug Resistance in *Mycobacterium tuberculosis.* Curr Top Microbiol Immunol. 2013;374:53-80. DOI: 10.1007/82_2012_279

[15] Gandhi NR, Nunn P, Dheda K, Schaaf HS, Zignol M, van Soolingen D, Jensen P, Jaime Bayona J. Multidrug-resistant and extensively drug-resistant tuberculosis: a threat to global control of tuberculosis. Lancet. 2010;375:1830-1843. DOI: 10.1016/S0140-6736(10)60410-2

[16] Kozińska M, Brzostek A, Krawiecka D, Rybczyńska M, Zwolska Z, Augustynowicz-Kopeć E. MDR, pre-XDR and XDR drug-resistant tuberculosis in Poland in 2000-2009. Pneumonol. Alergol. Pol. 2011;79,4: 278-287.

[17] Pontali E, Visca D, Centis R, D'Ambrosio L, Spanevello A, Migliori GB. Multi and extensively drug-resistant pulmonary tuberculosis: advances in diagnosis and management. Curr Opin Pulm Med. 2018;24(3):244-252. DOI: 10.1097/MCP.0000000000000477

[18] Perez-Lago L, Navarro Y, Garcia-de-Viedma D. Current knowledge and pending challenges in zoonosis caused by *Mycobacterium bovis*: a review. Res Vet Sci. 2014;97:94-100. DOI: 10.1016/j.rvsc.2013.11.008

[19] Kozińska M, Krajewska-Wędzina M, Augustynowicz-Kopeć E. *Mycobacterium caprae* - the first case of the human infection in Poland. Ann Agric Environ Med. 2020;27:151-153. DOI: 10.26444/aaem/108442

[20] Smith NH, Upton P. Naming spoligotype patterns for the RD9-deleted lineage of the *Mycobacterium tuberculosis* complex; www.Mbovis.org". Infect Genet Evol. 2012;12:873-876. 2012. DOI: 10.1016/j.meegid.2011.08.002

[21] Merker M, Kohl TA, Niemann S, Supply P. The evolution of strain typing in the *Mycobacterium tuberculosis* Complex. Adv Exp Med Biol. 2017;1019:43-78. DOI: 10.1007/978-3-319-64371-7_3

[22] Hermans PW, van Soolingen D, Dale JW, Schuitema AR, McAdam RA, Catty D, van Embden JD. Insertion element IS986 from *Mycobacterium tuberculosis*: a useful tool for diagnosis and epidemiology of tuberculosis. J Clin Microbiol. 1990 Sep;28(9):2051-2058.

[23] Kozińska M, Augustynowicz-Kopeć E. The incidence of tuberculosis transmission among family members and outside households. Pneumonol Alergol Pol. 2016;84:271-277. DOI: 10.5603/PiAP.2016.0034

[24] Kozińska M, Augustynowicz-Kopeć E. Drug Resistance and Population Structure of *Mycobacterium tuberculosis* Beijing Strains Isolated in Poland. Pol J Microbiol. 2015;64:399-401. DOI: 10.5604/17331331.1185243

[25] Nikolayevskyy V, Kranzer K, Niemann S, Drobniewski F. Whole genome sequencing of *Mycobacterium tuberculosis* for detection of recent transmission and tracing outbreaks: A systematic review. Tuberculosis (Edinb). 2016;98:77-85. DOI: 10.1016/j.tube.2016.02.009

[26] Supply P, Mazars E, Lesjean S, Vincent V, Gicquel B, Locht C. Variable human minisatellite-like regions in the *Mycobacterium tuberculosis* genome. Mol Microbiol. 2000;36(3):762-771. doi: 10.1046/j.1365-2958.2000.01905.x

[27] Roetzer A, Diel R, Kohl TA, Rückert C, Nübel U, Blom J, Wirth T, Jaenicke S, Schuback S, Rüsch-Gerdes S, Supply P, Kalinowski J, Niemann S. Whole Genome Sequencing versus traditional genotyping for investigation of a *Mycobacterium tuberculosis* outbreak: A Longitudinal Molecular Epidemiological Study. PLoS Med. 2013;10(2):e1001387. DOI: 10.1371/journal.pmed.1001387

[28] Animal Disease Notification System (ADNS) | Food Safety (europa.eu)

[29] Muñoz Mendoza M, Juan L d., Menéndez S, Ocampo A, Mourelo J, Sáez J L, Domínguez L, Gortázar C, García Marín J F, Balseiro A. Tuberculosis due to *Mycobacterium bovis* and *Mycobacterium caprae* in sheep. Vet J. 2012;191:267-269.

[30] Monreal L, Segura D, Segalés J, Garrido J M, Prades M. Diagnosis of *Mycobacterium bovis* infection in a mare. Vet Rec. 2001;149:712-714.

[31] van der Burgt GM, Crawshaw T, Foster AP, Denny D, Schock A. *Mycobacterium bovis* infection in dogs. Vet Rec. 2009;165:634.

[32] Krajewska-Wędzina M, Kozińska M, Radulski Ł, Lipiec M, Augustynowicz-Kopeć E, Weiner M, Szulowski K. Molecular characterisation of the *Mycobacterium bovis* causing bovine tuberculosis outbreaks in Poland. J Vet Res. 2020;64. DOI:10.2478/jvetres-2020-0003

[33] Sandoval Barron E, Swift B, Chantrey J, Christley R, Gardner R, Jewell Ch, McGrath I, Mitchell A, O'Cathail C, Prosser A, Ridout S, Sanchez-Cabezudo G, Smith N, Timofte D, Williams N, Bennett M. A study of tuberculosis in road traffic-killed badgers on the edge of the British bovine TB epidemic area. Sci Rep. 2018;8:17206. doi: 10.1038/s41598-018-35652-5

[34] Risco D, Martínez R, Bravo M, Fernández Llario P, Cerrato R, Garcia-Jiménez W L, Gonçalves P, García A, Barquero-Pérez O, Quesada A, de Mendoza JH. Nasal shedding of *Mycobacterium tuberculosis* in wild boar is related to generalised tuberculosis and concomitant infections. Vet Rec. 2019;185(20):629. doi: 10.1136/vr.105511.

[35] Krajewska-Wędzina M, Kozińska M, Augustynowicz-Kopeć E, Weiner M, Szulowski K. Molecular characterization of *Mycobacterium tuberculosis* complex strains isolated from cattle in Masovian Voivodeship in 2008-2012. Med Weter. 2018;74:133-138. DOI: 10.21521/mw.6060

[36] Krajewska M, Kozińska M, Zwolska Z, Lipiec M, Augustynowicz-Kopeć E, Szulowski K. Human as a source of tuberculosis for cattle. First evidence of transmission in Poland. Vet Microbiol 2012;159:269-271. DOI: 10.1016/j.vetmic.2012.04.001

[37] Krajewska, M. Characterization of *Mycobacterium bovis* strains isolated from animals in Poland [PhD thesis]. Pulawy, Poland: National Veterinary Research Institute, 2014.

[38] Krajewska-Wędzina M, Kozińska M, Orłowska B, Weiner M, Szulowski K, Augustynowicz-Kopeć E, Anusz K, Smith NH. Molecular characterization of *Mycobacterium caprae* strains isolated in Poland. Vet Rec 2018;182:1-6. DOI: 10.1136/vr.104363

[39] Krajewska M, Załuski M, Zabost A, Orłowska B, Augustynowicz-Kopeć E, Anusz K, Lipiec M, Weiner M, Szulowski K. Tuberculosis in antelopes in a zoo in Poland – Problem of Public Health. Pol J Microbiol 2015;4:405-407. DOI: 10.5604/17331331.1185242

[40] Krajewska-Wędzina M, Augustynowicz-Kopeć E, Weiner M, Orłowska B, Anusz K, Szulowski K. Tuberculosis in Polish zoos as health risk for humans. Health Prob Civil. 2017;11(4):233-238. DOI: 10.5114/hpc.2017.71892

[41] Krajewska-Wędzina M, Augustynowicz-Kopeć E, Weiner M, Szulowski K. Treatment in active tuberculosis in giraffe (*Giraffa camelopardalis*) in Silesian Zoo and its potential consequences for public health – Case report. Ann Agr Env Med. 2017. DOI: 10.26444/aaem/75685.

[42] Krajewska-Wędzina M, Olech W, Kozińska M, Augustynowicz-Kopeć E, Weiner M, Szulowski K. Bovine tuberculosis outbreak in farmed American bison (*Bison bison*) in Poland. Pol J Vet Sci 2017;20(4):819-821. DOI: 10.1515/pjvs-2017-0103

[43] Lipiec M, Radulski Ł, Szulowski K. A case of bovine tuberculosis in pigs in Poland – a country free from the disease. Ann Agr Env Med 2019;26:29-32. DOI: 10.26444/aaem/90979.

[44] Krajewska-Wędzina M, Didkowska A, Sridhara AA, Elahi R, Johnathan-Lee A, Radulski Ł, Lipiec M, Anusz K, Lyashchenko KP, Miller MA, Waters WR. Transboundary tuberculosis: Importation of alpacas infected with *Mycobacterium bovis* from the United Kingdom to Poland and potential for serodiagnostic assays in detecting tuberculin skin test false-negative animals. Transbound Emerg Dis. 2020, DOI: 10.1111/tbed.13471.

[45] Krajewska M, Welz M, Brewczyński P, Orłowska B, Anusz K. Bovine tuberculosis in the Bieszczady bison population. Życie Wet. 2014;89:148-115.

[46] Weiner M, Krajewska M, Zabost A, Augustynowicz-Kopeć E, Szulowski K. Analysis of resistance to antimycobacterial drugs among MTBC strains isolated from cattle in Poland as a threat to human health. Health Prob Civil. 2020;14(1):24-28. DOI: 10.5114/hpc.2020.92423

[47] Commission Decision 2009/342/EC as regards the declaration that certain administrative regions of Poland are officially free of zoonotic – bovine – leucosis and that Poland and Slovenia are officially free of bovine tuberculosis. 2009; Oj L 104, 24. 4. 2009:51-56.

[48] Orłowska B, Augustynowicz-Kopeć E, Krajewska M, Zabost A, Welz M, Kaczor S, Anusz K. *Mycobacterium caprae* transmission to free-living grey wolves (*Canis lupus*) in the Bieszczady Mountains in Southern Poland. Eur J Wildl Res. 2017;63:21. DOI: 10.1007/s10344-017-1079-4

[49] Welz M, Anusz K, Salwa A, Zaleska M, Bielecki W, Osińska B, Kaczor S, Kita Bovine tuberculosis in European bison in the Bieszczady region. Med Weter. 2005;61:441-444.

[50] Radulski Ł, Lipiec M, Krajewska-Wędzina M. Gruźlica bydlęca u zwierząt dzikich oraz wolno żyjących – badania laboratoryjne 2008-2018. Życie Weterynaryjne. 2019;94:51-53.

[51] Krajewska M, Lipiec M, Zabost A, Augustynowicz-Kopeć E, Szulowski K. Bovine tuberculosis in a wild boar (*Sus scrofa*) in Poland. J Wildl Dis. 2014;50,1001-1002. DOI: 10.7589/2013-07-187.

[52] Żurawski C, Lipiec M. Generalized tuberculosis in European bison. Med Weter 1997;53:90-92.

[53] Welz M. The epizootic situation among livestock and the wildlife in the area of Bieszczady Mountains considering the *Mycobacterium bovis* infections [PhD thesis]. Warsaw: University of Life Sciences, 2010.

[54] Lyashchenko KP, Sridhara AA, Johnathan-Lee A, Sikar-Gang A, Lambotte P, Esfandiari J, Bernitz N, Kerr TJ, Miller MA, Waters WR. Differential antigen recognition by serum antibodies from three bovid hosts of *Mycobacterium bovis* infection. Comp Immunol Microbiol Infect Dis. 2020;69:101424. DOI: 10.1016/j.cimid.2020.101424

[55] Goosen, WJ, Kerr TJ, Kleynhans L, Buss P, Cooper D, Warren RM, van Helden PD, Schröder B, Parsons S, Miller MA. The VetMAX™ *M. tuberculosis* complex PCR kit detects MTBC DNA in antemortem

and postmortem samples from white rhinoceros (*Ceratotherium simum*), African elephants (*Loxodonta africana*) and African buffaloes (*Syncerus caffer*). BMC Vet Res. 2020;16(1):2020. DOI: 10.1186/s12917-020-02438-9

[56] Miller MA, Finnegan M, Storms T, Garner M, Lyashchenko KP. Outbreak of mycobacterium tuberculosis in a herd of captive Asian elephants (*Elephas maximus*): antemortem diagnosis, treatment, and lessons learned. J Zoo Wildl Med. 2018;49(3):748-754. DOI: 10.1638/2017-0200.1.

[57] Didkowska A, Orłowska B, Witkowski L, Olbrych K., Brzezińska S, Augustynowicz-Kopeć E, Krajewska-Wędzina M, Bereznowski A, Bielecki W, Krzysiak M, Rakowska A, Olech W, Miller MA, Waters WR, Lyashchenko KP, Anusz, K. Biopsy and tracheobronchial aspirates as additional tools for the diagnosis of bovine tuberculosis in living European bison (*Bison bonasus*). Animals. 2020:10;2017. DOI: 10.3390/ani10112017

[58] DidkowskaA,Krajewska-WędzinaM, Bielecki W, Brzezińska S, Augustynowicz-Kopeć E, Olech W, Anusz K, Archana A.Sridhara AA, Johnathan-Lee A, Elahi R, Miller MA, Waters RW, Lyashchenko KP. Antibody responses in European bison (*Bison bonasus*) naturally infected with *Mycobacterium caprae*. Vet Microb. 2020. DOI: 10.1016/j.vetmic.2020.108952

[59] Cierpiał-Wolan M. Environmental protection in Podkarpackie voivodship in 2010-2012. Rzeszów, 2013, ISSN 1643-0271

[60] Żurawski C, Lipiec M. Infekcja *Mycobacterium bovis* u żubra. Med Weter. 1998;54:178-118.

[61] Kita J, Anusz K, Salwa A, Welz M, Orłowska B, Zalewska M. Bovine Tuberculosisin European Bison as Possible Zoonotic Impact in Poland. In: Lorenzo-Morales J, editor. Zoonosis 2012. p. 101-110.

[62] Krajewska M, Zabost A, Welz M, Lipiec M, Orłowska B, Anusz K, Brewczyński P, Augustynowicz-Kopeć E, Szulowski K, Bielecki W, Weiner M. Transmission of *Mycobacterium caprae* in a herd of European bison in the Bieszczady Mountains, Southern Poland. Eur J Wildl Res. 2015;61:429-433. DOI: 10.1007/s10344-015-0912-x

[63] Orłowska B, Krajewska-Wędzina M, Augustynowicz-Kopeć E, Kozińska M, Brzezińska S, Zabost A, Didkowska A, Welz M, Kaczor S, Żmuda P, Anusz K. Epidemiological characterization of *Mycobacterium caprae* strains isolated from wildlife in the Bieszczady Mountains, on the border of Southeast Poland. BMC Vet Res. 2020;16(1):362. DOI:10.1186/s12917-020-02581-3

[64] Czarnowski A. 1956. Gruźlica u sarny wywołana prątkiem typu ludzkiego. Med Weter. 1956;12:348.

[65] Krajewska M, Kozińska M, Orłowska B, Augustynowicz-Kopeć E, Anusz K, Welz M, Szulowski K. 2016. *Mycobacterium bovis* u żubra w Puszczy Boreckiej. In: Żubry w Krainie Dinozaurów, 8-9 September 2016, Bałtów, 2016, pp. 40-42.

[66] Didkowska A, Orłowska B, Krajewska-Wędzina M, Augustynowicz-Kopeć E, Brzezińska S, Żygowska M, Wiśniewski J, Kaczor S, Welz M, Olech W, Anusz K. Microbiological and molecular monitoring for bovine tuberculosis in the Polish population of European bison (*Bison bonasus*). Ann Agric Environ Med. 2020. DOI: 10.26444/aaem/130822

[67] Brewczyński P. The current activities for protection and breeding of European bison in Bieszczady. Europ Bison Conserv Newslett. 2013;6:123-132.

[68] Mazur J. Wisents in Forest District Stuposiany. Europ Bison Conserv Newslett. 2013;6:133-144.

[69] Smith NH, Upton P. Naming spoligotype patterns for the RD9-deleted lineage of the *Mycobacterium tuberculosis* complex; www.Mbovis.org. Infect Genet Evol. 2012;12:873-876. DOI: 10.1016/j.meegid.2011.08.002

[70] Kamerbeek J, Schouls L, Kolk A, van Agterveld M, van Soolingen D, Kuijper S, Bunschoten A, Molhuizen H, Shaw R, Goyal M, van Embden J. Simultaneous detection and strain differentiation of *Mycobacterium tuberculosis* for diagnosis and epidemiology. J Clin Microbiol. 1997;35:907-914. DOI: 10.1128/JCM.35.4.907-914.1997

[71] Alonso-Rodríguez N, Martínez-Lirola M, Herránz M, Sanchez-Benitez M, Barroso P. Evaluation of the new advanced 15-loci MIRU-VNTR genotyping tool in *Mycobacterium tuberculosis* molecular epidemiology studies. BMC Microbiol. 2008;24:8-34. DOI: 10.1186/1471-2180-8-34

Section 5

Other TB

Chapter 7

Humeral Artery Aneurysm Revealing a Rare Association between Tuberculosis and Behçet's Disease

Rabie Ayari, Ramy Triki, Youssef Mallat, Achraf Abdennadher, Khalil Amri, Raja Amri and Mohamed Ali Sbai

Abstract

The association of pulmonary tuberculosis and Behçet's disease revealed by an aneurysm of the humeral artery is exceptional with a complicated management. We report a case in which the two conditions occurred concomitantly with the vascular complication, apart from any use of immunosuppressive therapy, something that has never been reported in the literature. We report an extremely rare case of a spontaneous rupture of an aneurysm of the humeral artery of a 29-year-old woman, with no history. The patient underwent axillo-humeral bypass. Investigations concluded the diagnosis of Behçet's disease associated with pulmonary and lymph node tuberculosis. Anti-tuberculous chemotherapy followed by corticosteroids, immunosuppressants and colchicine have been administrated. Based on this observation, we insist on the necessity of searching the symptoms of Behçet's disease in the presence of arterial involvement when having a young patient. Therapeutic management must include medical treatment to control inflammation and limit the risk of recurrence. Endovascular or surgical treatment is necessary if the arterial involvement is threatening. The association with tuberculosis complicates management and requires close monitoring.

Keywords: Behçet's disease, tuberculosis, aneurysm, humeral artery, imaging

1. Introduction

Behçet's disease is an inflammatory, systemic vasculitis originally described by the Turkish dermatologist Hulusi Behçet in 1937. It is typically characterized by the combination of recurrent oral and genital aphtosis with ocular involvement. Vascular involvement in Behçet's disease occurs in 5 to 40% of cases depending on the series [1] and is associated with increased mortality [2]. Venous thrombosis is frequently observed, while arterial damage such as aneurysms, pseudoaneurysms, strictures and occlusions are less reported [2]. Arterial involvement in Behçet's disease mainly affects the aorta and pulmonary arteries. The humeral artery is extremely rarely affected.

The association of Behcet's disease and tuberculosis is rare and often correlated with the use of immunosuppressants.

To our knowledge, there is no reported case in the literature of spontaneous rupture of an aneurysm of the humeral artery of a patient having Behçet's disease and tuberculosis. Thus, we report the first case revealing this association.

2. Observation

We report the case of a 29-year-old woman with no history who presented with a right humeral mass of 50 mm x70mm evolving for 2 months and that has spontaneously ruptured. An axillo-humeral bypass using the basilic vein was made. The intraoperative findings were consistent with an aneurysm of the humeral artery. The axillary artery had an inflammatory aspect with a very thickened wall suggesting vasculitis, and the humeral artery downstream was narrowed.

Histological analysis of the resected lesions showed a non-specific panvascularitis.

Investigations carried out postoperatively revealed recurrent bipolar aphtosis associated with a 2-year history of inflammatory arthralgia and a 2-month history of claudication of the left upper limb. Physical examination showed a mouth ulcer on the inside of the lower lip (**Figure 1**), scars of genital ulcers, pseudo folliculitis. Peripheral pulses were present, symmetrical but weak in the left upper limb. Pathergy test was positive.

Arterial Doppler objectified the patency of the left axillary, subclavian, humeral, radial and ulnar arteries which were thin with damped and demodulated spectra. A thoraco-abdominopelvic CT angiography was performed in search of other vascular lesions. It showed a saccular aneurysm of the left subclavian artery, bronchiolar micronodules, thoracic and sub-diaphragmatic lymphadenopathies, the largest of which contained central necrosis and some of which were calcified. The diagnosis of Behçet's disease with mucocutaneous and arterial involvement, associated with tuberculosis was made. Anti-tuberculous chemotherapy followed by corticosteroids and immunosuppressants (cyclophosphamide and Azathioprine), as well as colchicine, has led to an uneventful recovery without recurrence.

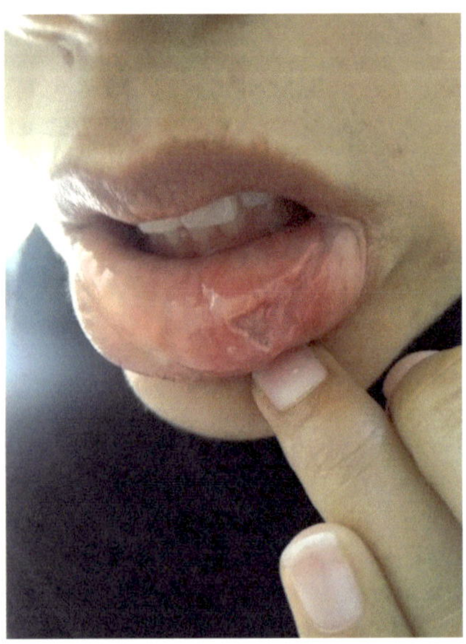

Figure 1.
Mouth ulcer on the inside of the lower lip.

3. Discussion

Tuberculosis is an infectious disease that presents a public health problem in developing countries where it is endemic [3]. Tuberculosis continues to be a major cause of morbidity and mortality. In developed countries, there has been an upsurge in the last decade, especially among HIV carriers, immigrant population and the elderly [4].

Tunisia is an intermediate-endemic country with a recorded incidence of 35/100,000 inhabitants in 2019 [5].

Extra-pulmonary tuberculosis accounts for 15 to 30% of all locations [6]. Ganglion, pleural, urogenital and bone sites are the most common [7].

Symptoms can be various depending on the location of the tuberculosis. General signs are often seen like weight loss, anorexia and fever, but none of them is specific.

The main cause of tuberculosis is *Mycobacterium tuberculosis*, a thin, slightly curved, aerobe bacillus. In comparison to other bacteria, M tuberculosis has a cell wall with a very high lipid content that resists staining by the usual Gram method. However, it accepts basic fuchsin dyes and is not easily discolorized even with acidalcohol; this resistance to decolorization by acid-alcohol is termed acid-fast [8]. M tuberculosis is transmitted via airborne droplet nuclei that are produced when persons with pulmonary or laryngeal tuberculosis cough, sneeze, speak, or sing [9].

Diagnosing active tuberculosis can be difficult. A chest X-ray and multiple sputum cultures for acid-fast bacilli are typically part of the initial evaluation [10].

Tuberculin testing is helpful, but it is not specific, and a negative test cannot exclude the diagnosis.

Cultures are slow but nevertheless remain the gold standard [11]: they allow the diagnosis to be confirmed and an antibiogram to be obtained. But they are rarely positive in paucibacillary tuberculosis.

The histopathological study is the key exam to prove the diagnosis of tuberculosis by objectifying the tuberculoid granuloma with caseous necrosis.

New diagnostic tools are available today such as Quantiféron.

Polymerase chain reaction should be used when having a positive direct examination, in order to distinguish the bacilli of the *Mycobacterium tuberculosis* from other atypical mycobacteria.

The treatment consists of six months of antituberculosis chemotherapy: rifampicin and isoniazid, initially supplemented by two months of pyrazinamide and ethambutol [12].

Behçet's disease is a rare but severely debilitating vasculitis. The manifestations are typically mucocutaneous with orogenital ulcers and skin lesions [13]. However, many other locations can be seen.

The symptoms are variable, which can explain the delay of the diagnosis in addition to the absence of specific blood test.

Behçet's disease occurs worldwide but clusters are found mainly along the 'silk road' with highest prevalence in Turkey, Japan and Iran, and lower prevalence in North American and northern European populations [14].

It affects people of all ages with a predilection for those aged from 20 to 40 years. Sex distribution is variable. The disease is usually severe in young adult men.

Behçet's disease can affect potentially all organ systems because of its propensity to affect all arteries and veins.

Oral and genital ulcers are the hallmarks of the disease, seen in up to 97% and 60–90% of patients, respectively [15].

The confirmation of diagnosis is based on appropriate clinical symptoms after exclusion of differential diagnoses.

Vascular involvement during Behçet's disease affects preferentially young males with a sex ratio M/F: 5/1 [16]. It is more frequent in the Middle East and Mediterranean countries. Our patient is a 36-year-old woman.

Most often, vascular involvement in Behçet's disease affects the venous system, usually in the form of thrombosis. Arterial involvement affects around 10% of patients and makes the severity of the disease [17]. The main arterial lesions are aneurysms, occlusions and more rarely arterial stenosis or diffuse aortitis [18].

Anatomopathological examination often shows active lesions made of an inflammatory infiltrate preferentially affecting the media, the adventitia and the surrounding of the vasa vasorum. It is associated with scar lesions with fibrous thickening in the media, the adventitia and the intima, all leading to the distension of the walls and the constitution of aneurysms or pseudo-aneurysms.

Aneurysms are by far one of the most serious and feared complications of Behçet's disease, especially the pulmonary location. All arteries can become aneurysmal, the main localizations are aortic, femoral, pulmonary, iliac, popliteal and subclavian (**Figure 2**). The other rarer arterial localizations are digestive, coronary, cerebral and upper limbs [19, 20]. In the series of Saadoun et al. [20], the arterial lesions of the lower limbs were more frequent than those of the upper limbs with respectively 51 cases vs. 5. Nevertheless, we should mention that the arterial localizations are willingly multiple in approximately 30% of the cases [19].

The rupture of the aneurysms of the humeral artery is dominated by traumatic causes (direct contusion, arteriography, arterial catheterization, blood gas, pulmonary biopsy, arterial bypass). In our case, the spontaneous rupture of the aneurysm of the humeral artery would probably be due to its chronicity and its voluminous size.

Ultrasound coupled with color and pulsed doppler, is the key exam and the first to request in front of a pulsatile mass, it allows the diagnosis to be made by showing an increase in the caliber of the artery.

Behçet's disease should be suspected when having a young patient with an arterial aneurysm of the upper limb. Clinical signs in favor should be sought by vigorous interrogation and examination, in order to avoid delayed diagnosis leading to serious complications (**Figure 3**).

The choice between endovascular or surgical treatment of aneurysmal lesions is not well codified, but it should ideally be made at a distance from the acute phase. However, many of these lesions are diagnosed at the stage of rupture or pre-rupture requiring emergency surgery which is the case of our patient.

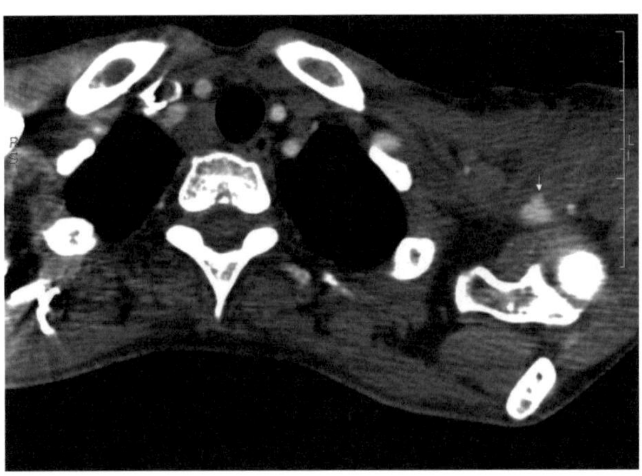

Figure 2.
Thoracic CT angiography showing a saccular aneurysm of the left subclavian artery (cross section).

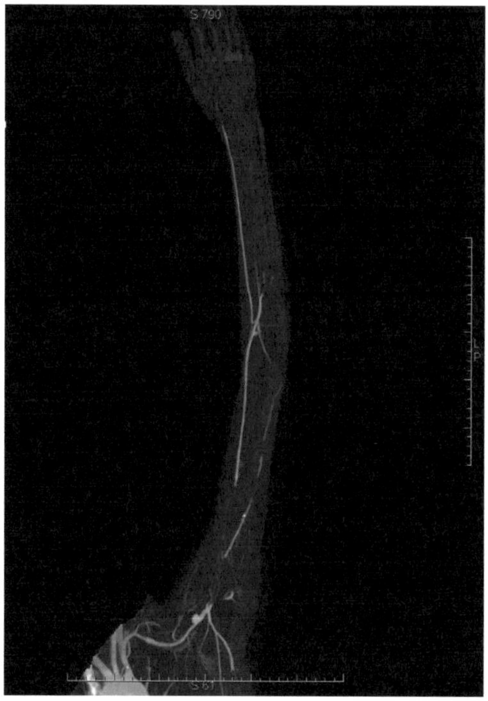

Figure 3.
CT angiography of the left upper limb showing: Fusiform aneurysm of the subclavian artery, occlusions of the left humeral artery of over 13 cm and occlusion of the lower two thirds of the left ulnar artery (reconstruction image).

Control of inflammation is essential, therefore high dose of corticosteroid therapy is recommended for serious arterial damage in combination with immunosuppressants [21].

The association of Behçet's disease with aneurysmal involvement to pulmonary tuberculosis is very rarely reported in the literature [22] and concerns the pulmonary arteries in most cases. The involvement of the humeral artery in the case of our patient is the first ever to be described in the literature in our knowledge.

The rare cases of association of Behçet's disease with other pathologies, notably infectious, such as pulmonary tuberculosis have been particularly observed when treatments such as TNF-a blockers, or other immunosuppressants were newly introduced. That was not the case in our observation which makes it special.

Apart from immunosuppressive therapy, Efthimiou et al. [23], explains this association by the disruption of the immune system induced by the disease itself, by genetic predisposition and by the ethnic factor which seems to be important in our case, since Tunisia is a tuberculosis endemic country. All of this highlights the value of an anti-bacillary prophylactic treatment in the case of immune diseases.

4. Conclusion

Arterial involvement during Behçet's disease is one of the main causes of mortality and morbidity. Aneurysms are the most common form, mainly affecting the aorta, the pulmonary and femoral arteries. The arteries of the upper limbs are rarely affected but they can inaugurate the disease and even be life-threatening in case of spontaneous or traumatic rupture. Hence the importance of suspecting Behçet's disease when facing any arterial aneurysm of the upper limb of a young patient.

Pulmonary or extra-pulmonary tuberculosis can be associated with Behçet's disease even outside the use of an immunosuppressive therapy. Here comes the utility to carry out screening tests or even to administer prophylactic treatment based on Isoniazid in endemic countries before any other medication.

Competing interests

Authors have declared that no competing interests exist.

Consent and ethical approval

As per university standard guideline, participant consent and ethical approval have been collected and preserved by the authors.

Author details

Rabie Ayari[1*], Ramy Triki[2], Youssef Mallat[1], Achraf Abdennadher[1], Khalil Amri[1], Raja Amri[3] and Mohamed Ali Sbai[2]

1 Department of Orthopedics and Trauma Surgery, Military Hospital of Instruction of Tunis, Tunisia

2 Plastic, Hand Surgery and Burns Department, Maamouri Hospital, Nabeul, Tunisia

3 Internal Medicine Department, Maamouri Hospital, Nabeul, Tunisia

*Address all correspondence to: rabie.ayari@hotmail.com

References

[1] Düzgün N, Ateş A, Aydintuğ OT, Demir O, Olmez U. Characteristics of vascular involvement in Behçet's disease. Scand J Rheumatol. févr 2006;35(1):65-68.

[2] Koç Y, Güllü I, Akpek G, Akpolat T, Kansu E, Kiraz S, et al. Vascular involvement in Behçet's disease. J Rheumatol. mars 1992;19(3):402-410.

[3] Ebelin M, Prosper M, Elfekih S. Tuberculose du poignet et de la main. In : Allieuy, dir. Infections de la main. Paris : EPS ;1998. p. 45-50.

[4] Shah MA, Shah I. Wrist swelling. Is it tuberculosis? J Family Med Prim Care. 2017;6(4):865-866.

[5] WHO. Global Tuberculosis Report 2019.

[6] Hopewell PC. A clinical view of tuberculosis. Imaging of tuberculosis and craniospinal tuberculosis. Radiol Clin North Am. 1995;33:641-653.

[7] Sharif HS, Morgan JL, Ai Shahed MS, Ai Thagafi MY. Role of CT and MRI in the management of tuberculous spondylitis. Imaging of tuberculosis and craniospinal tuberculosis. Radiol Clin North Am. 1995;33:787-804.

[8] Adkinson NF, Bennett JE, Douglas RG, Mandell GL (2010). Mandell, Douglas, and Bennett's principles and practice of infectious diseases (7th ed.). Philadelphia, PA: Churchill Livingstone/Elsevier. p. Chapter 250.

[9] Centers for Disease Control and Prevention. Guidelines for preventing the transmission of Mycobacterium tuberculosis in health-care facilities. MMWR Morb Mortal Wkly Rep 1994; 43(RR-13):1-132.

[10] Escalante P (June 2009). "In the clinic. Tuberculosis" Annals of Internal Medicine. 150 (11): ITC61-614, quiz ITV616.

[11] Truffot-Pernot C, Veziris N. Les tests bactériologiques de latuberculose maladie : standards et perspectives. Rev Mal Respir2011;28:1034—1047

[12] National Institute for Health and Clinical Excellence. Tuberculosis: clinical diagnosis and management of tuberculosis, and measures for its prevention and control. London: NICE, 2006.

[13] Behcet H. Uber rezidiveren deaphthose durch ein virus verursachte Geschwure am Mund, am Auge, und an den Genitalien. Dermatol Wochenschr. 1937;105:1152-1157.

[14] Yurdakul S, Günaydin I, Tüzün Y, Tankurt N, Pazarli H, Ozyazgan Y, Yazici H. The prevalence of Behçet's syndrome in a rural area in northern Turkey. J Rheumatol. 1988; 15(5):820-822.

[15] Nair JR, Moots RJ. Behcet's disease. Clin Med (Lond). 2017 Feb; 17(1): 71-77.

[16] Ko GY, Byun JY, Choi BG, Cho SH. The vascular manifestations of Behçet's disease: angiographic and CT findings. BJR. déc 2000;73(876):1270-1274.

[17] Park JH, Chung JW, Joh JH, Song SY, Shin SJ, Chung KS, et al. Aortic and Arterial Aneurysms in Behçet Disease: Management with Stent-Grafts—Initial Experience. Radiology. sept 2001;220(3):745-750.

[18] Lê Thi Huong D, Wechsler B, Papo T, Piette JC, Bletry O, Vitoux JM, et al. Arterial lesions in Behçet's disease. A study in 25 patients. J Rheumatol. nov 1995;22(11):2103-2113.

[19] Desbois A-C, Wechsler B, Cacoub P, Saadoun D. Atteintes aortiques inflammatoires associées à la maladie de Behçet. La Revue de Médecine Interne. avr 2016;37(4):230-238.

[20] Saadoun D, Asli B, Wechsler B, Houman H, Geri G, Desseaux K, et al. Long-Term Outcome of Arterial Lesions in Behçet Disease: A Series of 101 Patients. Medicine. janv 2012;91(1):18-24.

[21] Desbois A-C, Wechsler B, Cacoub P, Saadoun D. [Aortic inflammatory lesions in Behçet's disease]. Rev Med Interne. avr 2016;37(4):230-238.

[22] M. Haouel, C. Kooli, S. Azzabi, E. Cherif, Z. Kaoueche, L. Ben Hassine, N. Khalfallah ; La maladie de Behcet compliqué par un anévrysme de l'artère pulmonaire : A propos de 4 cas ; la société fransaised ethumatologie ; Ma.76 ; 2008.

[23] J. Efthimiou, P. E. Hay, S. G. Spiro* And D. J. Lane; Pulmonary tuberculosis in behçet's syndrome. Br. J. Dis. Chest (1988) 82, 300OME.